Pipelined Multiprocessor System-on-Chip for Multimedia

Haris Javaid · Sri Parameswaran

Pipelined Multiprocessor System-on-Chip for Multimedia

Springer

Haris Javaid
Sri Parameswaran
School of Computer Science
and Engineering
University of New South Wales
Kensington, NSW
Australia

ISBN 978-3-319-01112-7 ISBN 978-3-319-01113-4 (eBook)
DOI 10.1007/978-3-319-01113-4
Springer Cham Heidelberg New York Dordrecht London

Library of Congress Control Number: 2013948377

Printed on acid-free paper

Springer is part of Springer Science+Business Media (www.springer.com)

Contents

Chapter 1
Introduction

Day to day computing has moved from mainframe to personal to ubiquitous computing over the last several decades [1]. Ubiquitous computing is almost imperceptible and yet is everywhere around us, enabled by the proliferation of embedded systems. An embedded system is a hardware-software computer system, designed to perform specific tasks (unlike a general-purpose system) and is typically embedded within a larger system or device. Common examples of embedded systems include digital watches, traffic controllers, mobile phones, music/video players, tablets, health monitors and modern cars.

The evolution of embedded systems has been rapid and their market is growing at a staggering rate. In a report published by the International Data Corporation in 2011 [2], 5.4 billion embedded systems were shipped in 2010 and 8.8 billion are expected in 2015. Furthermore, 7.5 billion embedded processors were used in 2010 and 14.5 billion will most likely be required in 2015. Embedded systems' market includes a diverse set of industries spanning automotive, communication, consumer, energy, healthcare, industrial and transportation. The communication and consumer industries accounted for 48 % of the revenue of the embedded systems' market in 2010 [2], signifying user demands and expectations on consumer devices such as mobile phones, personal digital assistants, digital cameras, digital TVs and gaming consoles.

Multimedia is a combination of diverse content forms such as text, audio, video, image and animation to provide information or entertainment to users. It is at the backbone of consumer products and is considered the fastest growing class of embedded applications [3]. Users expect multimedia content to be accessible virtually from everywhere through portable devices. For example, a mobile phone is expected to record high definition video and then upload it to a social networking website. Another example is that users expect set-top boxes to provide digital television, internet access, gaming experience, in-home entertainment and home automation. The number of mobile phones has increased from 12.4 million to approximately 4.6 billion, and internet users have grown from 3 million to almost a quarter of the earth's population during the last two decades [4]. Therefore, analysis, design and

H. Javaid and S. Parameswaran, *Pipelined Multiprocessor System-on-Chip for Multimedia*, DOI: 10.1007/978-3-319-01113-4_1,

implementation of advanced and complex multimedia embedded devices has become an active research area in both academia and industry.

This monograph explores implementation of multimedia applications on a pipelined MultiProcessor System on Chip (MPSoC) where the processors are divided into stages, and are connected in a pipeline. Application Specific Instruction set Processors (ASIPs) [5] are used so that their customisation can be used to balance the workload across stages of the pipelined MPSoC, improving the utilisation of the processors for high performance, reduced area footprint and low power consumption. The aim of this monograph is to optimise such a pipelined MPSoC for area footprint and energy consumption under performance constraints by utilising design-time and run-time optimisations. This chapter of the monograph entails an overview of trends and challenges in multimedia applications and embedded architectures, starting from low resolution video processing on uniprocessor systems to high definition video processing on (heterogeneous) multiprocessor systems.

1.1 Multimedia Applications

Multimedia has widespread application in embedded devices [6] in the form of:

- *Digital Audio*: audio recording, audio playback, voice calls/conferencing, etc.
- *Digital Image*: photography, image processing, image pre-/post-processing, etc.
- *Digital Video*: video calls/conferencing, video recording, video playback, digital TV, etc.
- *Display*: brightness and contrast adjustment, up-/down-scaling, etc.
- *Games*: game processing, rendering, shading, etc.

Multimedia applications have seen a radical increase in their complexity over the last two decades, driven by user expectations on better quality/experience, interactive displays, high definition content, 3D content, longer playback time, etc. Video resolutions have increased from Quarter Common Interface Format (QCIF, 176×144 pixels) to Standard Definition (SD, 720×480 pixels) to High Definition (HD, 1920×1080 pixels). These resolutions are expected to further increase to Ultra High Definition TV and Realistic TV, resulting in approximately $1000\times$ increase in resolution complexity [6, 7] relative to MPEG-4 QCIF. Although high resolutions are targeted for high-end devices, recent prototypes from Nokia have demonstrated 3D video decoding on tablets [8, 9]. In addition to video resolution, video codec complexity has dramatically increased to improve compression efficiency. Since the introduction of MPEG-1, video coding standards have evolved to H.264 [10] and Multiview Video Coding [11]. H.264 doubles the compression efficiency compared to previous standards [12] at the cost of $10\times$ additional computational complexity [13]. A recent study by Meehan et al. [7] anticipated that the overall complexity of video coding standards will double every two years. Besides, multimedia applications are expected to support different video formats and multiple video coding standards due to extreme competition in consumer devices' market.

High-end applications like user interfaces, video conferencing/calls, video recording and internet video streaming require better video quality, higher video resolutions and lower compression rates, and hence consume significant amounts of energy due to their high computational complexity. These applications are executed on portable devices like personal navigation devices, personal multimedia players, mobile internet devices and netbooks that are powered by batteries. As a result, embedded multimedia devices are anticipated to perform more than 100 Giga operations per second with power budgets of approximately 200 mW [7, 14]. Therefore, as multimedia moves towards 3D content at higher resolutions with multiple standards to live up to user demands and expectations, embedded multimedia hardware needs to be a flexible, computationally capable platform while being low power to run off a standard mobile battery. The next section describes the evolution of multimedia architectures.

1.2 Multimedia Architectures

Hardware architectures for multimedia have evolved significantly over the years, starting from Application Specific Integrated Circuits to Digital Signal Processors to Application Specific Instruction set Processors to Multiprocessor System-on-Chips.

1.2.1 Application Specific Integrated Circuits

Application Specific Integrated Circuits (ASICs) are integrated circuits designed and optimised for a specific application. ASIC designs are described in a Hardware Description Language (HDL) like VHDL and Verilog, which can then be simulated and synthesised by Electronic Design Automation (EDA) tools such as Mentor Graphics' ModelSim [15], Synopsys' Design Compiler [16] and Cadence's Virtuoso Platform [17]. ASICs provide high performance under tight area footprint and energy consumption budgets because of the highly optimised hardware. However, they provide a pure hardware solution which involves high design effort and lacks flexibility, and thus are increasingly becoming unattractive. Inflexibility and non-programmability of ASICs mean that they cannot be used for applications other than the ones for which they were initially designed. Therefore, ASICs need to be redesigned to support product upgrades which not only lengthens time-to-design and time-to-market of the product but also incurs significant Non-Recurring Engineering (NRE) costs. NRE costs are growing steadily with continuous technology scaling [4] which will make design reuse necessary, an attribute lacking in ASICs. Furthermore, support of multiple applications in a single ASIC will incur high design efforts due to the amplified design complexity. Therefore, programmable platforms turn out to be an attractive option for multimedia devices.

1.2.2 General Purpose Processors

General Purpose Processors (GPPs) offer a pure software solution that facilitates short time-to-design and time-to-market through code reuse, and allow easy product upgrades and fixes. Furthermore, programmability of GPPs helps longer time-in-market, and thus reduces NRE costs. Since GPPs cannot be optimised for specific applications, they offer far less performance and consume far more energy than ASICs. The quantitative analysis in [18] reported a difference of at least five orders of magnitude in energy efficiency[1] and area efficiency[2] between ASICs and GPPs.

1.2.3 Digital Signal Processors

Digital Signal Processors (DSPs), replacing GPPs, are domain-specific processors customised to efficiently execute applications from a certain domain. DSPs provide better energy and area efficiencies than GPPs [4, 6, 19] due to domain-specific instructions, multiple domain-specific functional units and exploitation of instruction- and data-level parallelisms. A typical DSP designed for multimedia applications will contain Multiply Accumulate (MAC), Fast Fourier Transform (FFT), Fused Multiply Add (FMA), etc. domain-specific functional units and associated instructions [20]. The Very Long Instruction Word (VLIW) technique allows a DSP to execute several operations in parallel, and the compiler is responsible for encapsulation of multiple operations in a single instruction. On the other hand, the Single Instruction Multiple Data (SIMD) technique allows an instruction to execute an operation on multiple data in parallel. VLIW and SIMD allow DSPs to exploit instruction- and data-level parallelism available in multimedia applications [3, 21].

Commercial DSPs for multimedia include Texas Instruments' C6000 series and DaVinci [20], FreeScale's StarCore [22], Analog Devices' SHARC, SigmaDSP and ADSP series [23], and NXP Semiconductor's TriMedia [24]. DSPs are also used as coprocessors with GPPs where GPPs offload domain-specific, computationally intensive functions to DSPs. For example, ConnX Vectra DSP coprocessor [25] is used with Tensilica's Xtensa processors [26] to perform fixed-point arithmetic for wireless communication applications. DSPs significantly improve energy and area efficiencies of GPPs while still being flexible and programmable. However, they

---

[1] Measured in mW/Million Operations Per Second.

[2] Measured in Million Operations Per Second/mm^2.

do not provide the best energy efficiency because they exploit a limited amount of parallelism and their performance is constrained by memory bandwidth [27–30].

1.2.4 Application Specific Instruction Set Processors

Application Specific Instruction Set Processors (ASIPs) [5, 31] emerged as an attractive platform to ASICs, GPPs and DSPs. ASIPs are highly customised processors with domain-specific hardware accelerators. These hardware accelerators are integrated with the processor pipeline and are accessible through custom instructions. Therefore, ASIPs provide better energy and area efficiencies than DSPs and GPPs [28, 32, 33] while retaining flexibility and programmability to support product upgrades and fixes with short time-to-design and time-to-market. The programmability feature (such as pipeline control, register file, etc.) of ASIPs results in a larger area footprint than ASICs, however technology scaling has subdued this shortcoming of ASIPs by making billions of transistors available on a single chip [34].

ASIPs provide numerous customisations, categorised into custom instructions, inclusion/exclusion of optimised domain-specific blocks, and parametrisable options [4, 35]. Custom instructions typically exploit the techniques of SIMD, VLIW and fused operations. Examples of optimised domain-specific blocks include multipliers, MAC units, Floating Point (FP) units and DSP coprocessors. Parametrisable options include pipeline depth, register file size, number of load-store units, local memory interface width, instruction and data caches, etc. An ASIP can be extremely tailored to an application due to the availability of such a diverse set of customisations, and hence provides the best tradeoff between area efficiency, energy efficiency, flexibility and programmability for multimedia applications [36–39]. Several commercial ASIP platforms are available from Tensilica [26], ARC International [40], CoWare [41], MIPS [42] and Target Compiler Technologies [43].

The design effort of an ASIP is extremely large because it not only involves design and verification of ASIP architecture but also the construction of the associated software tools such as assembler, compiler, debugger and instruction set simulator. However, several high-level ASIP frameworks have been developed over recent years to lower the design and verification efforts, and hence shorten time-to-design and time-to-market. These frameworks can be categorised as (inspired from [4]):

- *Specification based frameworks* [41, 43–45]: These frameworks let a designer develop an ASIP from scratch through specification of its Instruction Set Architecture (ISA) in an Architecture Description Language (ADL). Automatic generators are then used to create both the hardware model of the ASIP in HDL and corresponding software tool-chain.
- *Base processor based frameworks* [26, 40, 46]: These frameworks allow designers to develop an ASIP from a pre-designed and pre-verified configurable base processor. Designers can add functional units and custom instructions, and parametrise hardware blocks. Like specification based frameworks, the hardware model and

associated tool-chain is automatically generated. Furthermore, analysis tools are provided to automatically analyse applications and generate domain-specific hardware accelerators and associated custom instructions [47].

Recently, ASIPs have been coupled with Field Programmable Gate Array (FPGA) technology to create so-called reconfigurable processors. Like ASIPs, reconfigurable processors contain custom instructions; however, the corresponding hardware accelerators are implemented in the reconfigurable region which is integrated with the processor pipeline. The reconfigurable region is time-multiplexed among hardware accelerators to reduce area footprint when an ASIP does not use most of its custom instructions simultaneously. Reconfigurable processors further enhance the flexibility and programmability of ASIPs where both the hardware (through FPGA reconfiguration) and software (through code modification) can be modified. However, this increased flexibility comes at the cost of increased area footprint and power consumption of the FPGA fabric and its reconfiguration. Some examples of reconfigurable processors include MOLEN [48], WARP [49], RISPP [50], NIOS [51], eMIPS [52] and Stretch series [53], with detailed surveys in [54, 55].

1.2.5 Multiprocessor System-on-Chips

From GPPs to ASIPs, performance improvements were mostly due to exploitation of instruction- and data-level parallelisms, higher clock frequencies and technology scaling. Instruction- and data-level parallelisms did not scale well with the increase in complexity of multimedia applications, and hence single ASIP systems could not handle complexity of current multimedia [57–59]. Higher frequencies significantly increased dynamic power consumption while technology scaling increased leakage power consumption due to smaller transistor dimensions and reduced threshold voltages, increasing power densities and thus hitting the power wall [60–62]. Figure 1.1 illustrates that uniprocessor systems' clock frequencies (marked as "clock speed") and instruction-level parallelism capabilities (marked as "perf/clock (ILP)") have levelled off in the recent years. Therefore, rather than using a single complex, power inefficient processor, academia and industry went to explore the area of multiple, small, power efficient processors [62–64].

Continuous technology scaling (that is, 90 to 65 to 45 nm) has made billion of transistors available on a single chip to be exploited by System-on-Chip (SoC) technology to place multiple components on a single chip. A recent report from the International Data Corporation [2] noticed that SoCs will constitute the largest portion of embedded systems' market revenues, as shown in Fig. 1.2. The SoC technology has evolved over the years to fabricate MultiProcessor System-on-Chip (MPSoC) by putting together multiple processing elements, memory hierarchy, I/O components and an on-chip interconnect. Recent consumer products are believed to have up to ten processing elements in the form of MPSoCs [65]; for example, Apple's iPhone 5 has two processors while Samsung's Galaxy S III has four processors in the main

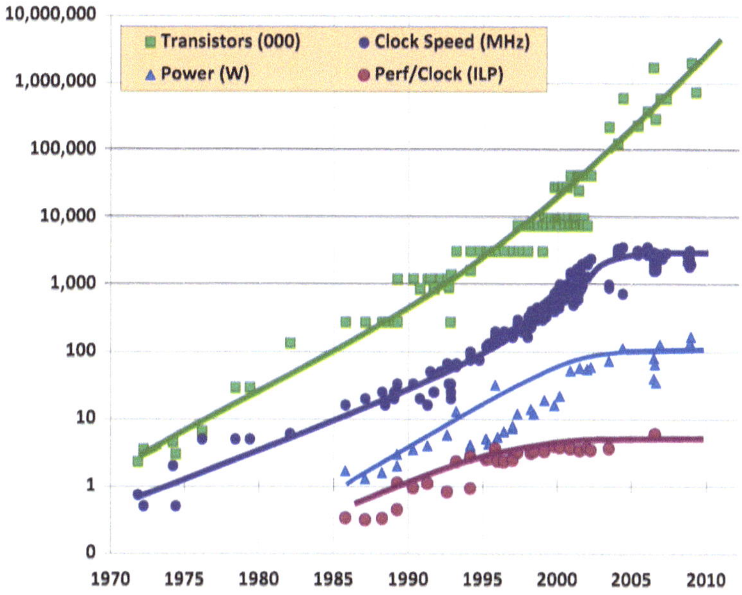

Fig. 1.1 Industry's move towards multiprocessor systems. Courtesy of Herb Sutter, sourced from [56]

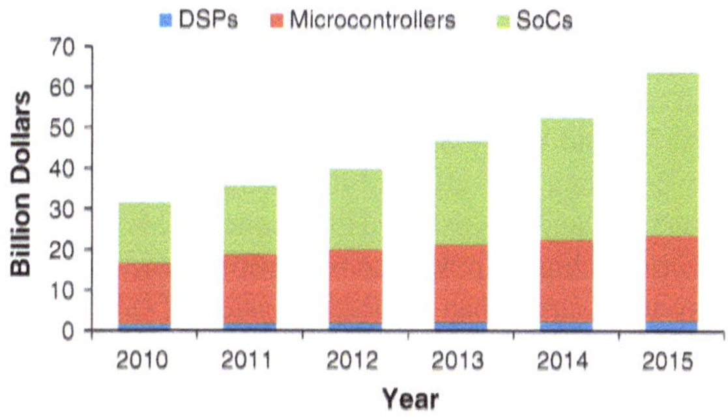

Fig. 1.2 Share of DSPs, microcontrollers and SoCs in revenue of embedded systems' market (data from [2])

(control) MPSoC. Futurists are expecting consumer products to contain MPSoCs with hundred processing elements in near future [62–64]. The International Technology Roadmap for Semiconductors (ITRS) has envisioned MPSoCs to contain even thousand processing elements by 2020, as depicted in Fig. 1.3. Therefore, MPSoCs have become a mainstream embedded platform for current multimedia applications [66]. In general, MPSoCs:

- can execute multiple applications with higher performance through exploitation of Task-Level Parallelism (TLP);

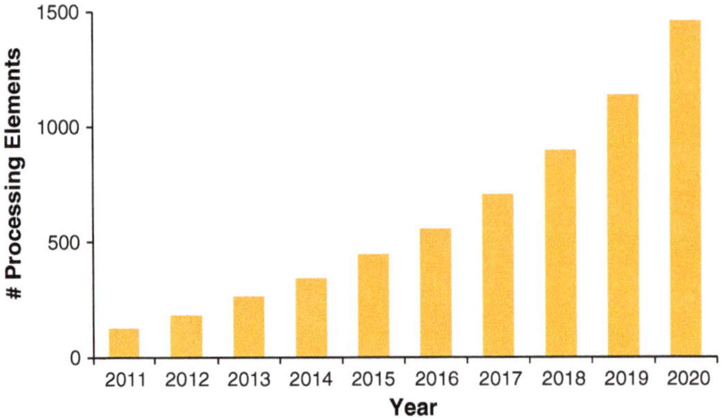

Fig. 1.3 Design trends of MPSoCs (data from [67])

- can consume less power by switching off idle processing elements;
- can be more reliable by sparing some processing elements for redundancy; and,
- can be scalable by the addition of more processing elements.

MPSoCs are broadly categorised as homogeneous or heterogeneous. Homogeneous MPSoCs are Symmetric MultiProcessing (SMP) systems where identical processing elements are used. For example, ARM's MPCore [68] contains four identical ARM11 processors with same Instruction Set Architecture (ISA), connected to a shared memory. Other notable examples are Stanford's Imagine [69], Tilera's TilePro64 [70] and Intel's Single-chip Cloud Computer (SCC) [71]. These MPSoCs typically contain a fast, efficient interconnect and an operating system to manage application tasks and processors. Homogeneous MPSoCs are scalable, have larger area footprint and higher power consumption; hence, are more suitable for general-purpose systems rather than embedded systems [72].

Heterogeneous MPSoCs are Asymmetric MultiProcessing (AMP) systems made up of architecturally different processing elements such as programmable processors (GPPs), application-specific processing elements (ASIPs, ASICs) and domain-specific (co) processors (DSPs), typically connected through a custom-designed interconnect. In such an architecture, processing elements are matched to the requirements of application's task(s), and hence heterogeneous MPSoCs provide high performance under tight area and power budgets. Several researches have shown that heterogeneous MPSoCs outperform their homogeneous counterparts [73–75], especially in multimedia [57, 58, 76, 77]. Commercially available heterogeneous MPSoCs for multimedia include Sony, Toshiba and IBM's CELL [78], Intel's IXP [79], NXP Semiconductor's Nexperia [80], Texas Instrument's OMAP [81] and STMicroelectronic's Nomadik [82].

Multimedia architectures have come a long way from ASICs to heterogeneous MPSoCs, and a figurative comparison is provided in Fig. 1.4. Heterogeneous MPSoCs have become an attractive platform for multimedia applications because:

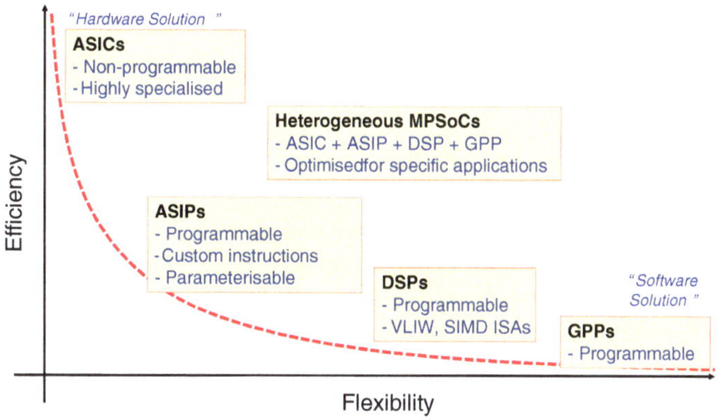

Fig. 1.4 Comparison of multimedia architectures (inspired from [6])

- Multimedia applications are heterogeneous in nature, that is, the type of computation, access patterns, memory bandwidth and workload profiles vary across tasks of a single application. For example, motion estimation in H.264 performs correlation on macroblocks and is highly data-dependent. On the other hand, discrete cosine transform performs large number of multiplications and additions with regular access pattern. Therefore, heterogeneous MPSoCs use customised processing elements to match the computational requirements of individual tasks.
- Customised processing elements typically result in lower area footprint and lower power consumption. Further power reductions can be achieved by switching off the idle processing elements through clock-gating, power-gating and Dynamic Voltage and Frequency Scaling (DVFS). Hence, heterogeneous MPSoCs can deliver the required performance under tight area and power budgets.
- Increasing complexities of multimedia applications (HD video, 3D video, etc.) can be addressed with the further addition of customised processing elements in the heterogeneous MPSoC.
- Heterogeneous MPSoCs built from domain-specific and application-specific processors (DSPs and ASIPs) can support multiple multimedia standards on the same platform through software, while still being able to deliver required performance under area and power budgets due to optimised processors.
- Domain-specific and application-specific processors based heterogeneous MPSoCs can also support multimedia features' fixes and updates, multimedia standard's upgrade and product upgrade through software modifications.

1.3 Challenges in Multimedia Heterogeneous MPSoCs

The advantages of heterogeneous MPSoCs come at a cost. Most importantly, their design becomes very complex due to the presence of a large number of architectural and programming options. Consumer market factors such as quick deployment, low prices, etc. put further pressure on the design of heterogeneous MPSoCs. This section describes these challenges and the motivation behind this monograph.

Time-to-Design and Time-to-Market

Consumer demands have forced semiconductor companies to regularly introduce and upgrade their products. For example, mobile phone companies have to release new models with innovative functionalities (such as face detection, higher pixel cameras, etc.) every 6 months or so to sustain their customer base. Not only this, companies have to release several variants of a mobile phone to capture diverse user expectations, and hence survive the competition. These market factors have resulted in shorter time-to-design and time-to-market for heterogeneous MPSoCs, indicating the need for comprehensive design automation frameworks.

Product Prices

Design of complex heterogeneous MPSoCs under short time-to-design and time-to-market constraints requires a company to invest in large, talented design teams. However, such investments mean that prices of products will increase which is unacceptable in consumer markets as users always prefer to buy state-of-the-art technology at the cheapest price. Design automation techniques can automatically run cumbersome phases of the design cycle with little or no intervention from a designer, and hence shortens time-to-design and time-to-market, which reduces product prices. Flake et al. [83] reports that a company with comprehensive design automation framework(s) is more likely to compete in consumer market by providing cheap yet innovative products.

Design Complexity

The design space of heterogeneous MPSoCs explodes due to the presence of diverse options such as processing elements, memory hierarchies, communication infrastructure and application/programming models. For example, should a heterogeneous MPSoC use ASIPs or DSPs or both, and how many of each type? In communication infrastructure, for example, a designer needs to choose from point-to-point buffers, shared memory buffers and their sizes, and Network-on-Chip (NoC). Choices in memory hierarchy include number of cache levels, configuration of caches and sizes of local and shared memories. Last but not least, should the heterogeneous MPSoC use the Kahn Process Network (KPN) [84], Synchronous Data Flow (SDF) [85] or stream model [86] for applications and OpenMP or Message Passing Interface

(MPI) as its programming model? Exploration of such a diverse design space cannot be done manually, and hence requires cleverly designed exploration techniques. Furthermore, design space exploration should be fast and implementable as part of the design automation framework(s).

Flexibility and Scalability

The heterogeneous MPSoC should be flexible enough to allow implementation of multiple multimedia standards so that several variants of a product can be quickly deployed. Furthermore, it should allow quick product fixes and upgrades after deployment. These requirements indicate the use of programmable processing elements like DSPs and ASIPs as the building blocks of a heterogeneous MPSoC. Design-time scalability implies that the MPSoC should allow easy addition of components in future to handle increasing complexity of next generation multimedia without major redesign effort.

Performance, Area and Energy Constraints

Multimedia applications often have performance constraints such as 30 fps for a video encoder that have to be met by heterogeneous MPSoCs. In addition, these MPSoCs are deployed in embedded devices running off standard batteries, and hence favour smallest possible area footprint and lowest possible power consumption. Design space exploration, as explained above, has to be performed to choose the right number and types of processing elements, cache configurations, memory sizes, type of low-power technique, etc. under performance, area and power constraints.

Adaptability

Computational requirements of a multimedia content changes with time, requiring multimedia applications and architectures to adapt accordingly at run-time. For example, a video encoder might be inputted with a video that contains low motion and then high motion. High-motion video frames require significantly more computation than low-motion video frames. Hence, a heterogeneous MPSoC should adapt its resource (processing elements, memory, etc.) utilisation at run-time based on current workload rather than operating under worst-case (that is, all the resources are active) at all times. Such an adaptation is necessary for ultra low-power operation of heterogeneous MPSoCs to increase battery lives in portable devices. Run-time management techniques should be used to manage resources in a heterogeneous MPSoC so that it always operates with the lowest possible power consumption.

1.4 Research Aims and Contributions

This monograph aims to address the above mentioned challenges, that can be condensed into the following three research problems:

- Selection of a suitable multimedia heterogeneous MPSoC platform;
- Design space exploration of the selected platform as part of design automation; and,
- Support for run-time adaptability in the selected platform.

Selection of an implementation platform is typical of platform-based design methodology [87] to keep the design complexity and design space of MPSoCs tractable. This monograph uses the paradigm of pipelined MPSoCs as the multimedia platform. A pipelined MPSoC is a system where processors are connected in a pipeline [88–92]. It is divided into several stages where each stage contains one or more processors. Communication between the stages typically occurs through point-to-point FIFO buffers. Each processor has separate instruction and data caches that are connected to its local memory. In addition to local memories, shared memory could be used where common data need to be shared among processors within a stage or across different stages. Figure 1.5 shows a typical four stage pipelined MPSoC. Pipelined MPSoCs have emerged as an attractive platform for multimedia [88–92] and offer several advantages that are summarised below:

- The data-flow nature of multimedia applications favours the topology of pipelined MPSoCs [93–96]. Multimedia applications are characterised by several sub-kernels which are executed repeatedly on an input data stream. For example, an MP3 encoder contains the following sub-kernels: Reading input file (R); Polyphase Filtering (PF); Transform and Quantisation (TQ); and, Entropy Coding and Writing output file (EC/W). These sub-kernels can be mapped to the four stages of the pipelined MPSoC shown in Fig. 1.5. While processor P2.1 will be in its ith iteration, processor P1.1 will be in its $(i+1)$th iteration, thereby allowing pipelined

Fig. 1.5 A typical pipelined MPSoC (memories are not shown for the sake of simplicity)

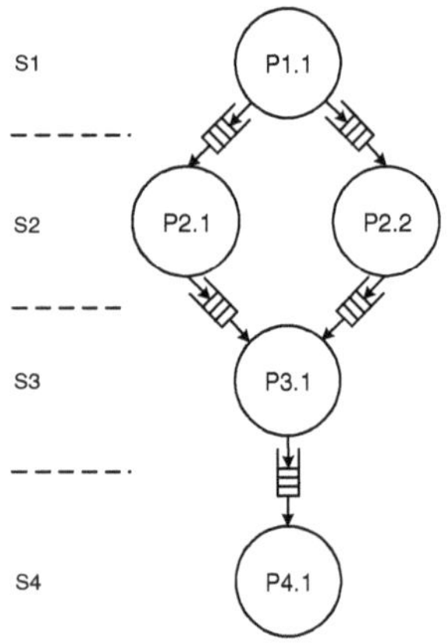

execution of the sub-kernels. Thus, these sub-kernels operate on different data units of the input stream and the incoming data streams through the stages of the pipelined MPSoC, enabling pipelined execution for high performance.

- Application Specific Instruction set Processors (ASIPs) are used as the processing elements which allow extreme customisation to match processors to sub-kernels, and thus deliver high performance with smaller area footprint and lower power consumption. A number of researches have illustrated the usefulness of ASIPs in multimedia MPSoCs [58, 65, 77, 97, 98]. ASIPs come with high-level frameworks that enable (semi-) automatic customisation [31, 47], reducing time-to-design and time-to-market of pipelined MPSoCs. Furthermore, ASIPs make pipelined MPSoCs flexible and scalable to support product upgrades through software.
- The point-to-point FIFO buffers allow communication at a much higher bandwidth compared to a shared bus and provide blocking read and write operations to allow synchronisation between processors. In addition, where FIFO buffers might have unacceptable area footprint, shared memories could be used for data communication [59].

The selection of pipelined MPSoCs as the multimedia platform limits the design space to be tractable, yet provides a high performance, flexible, explorable and customisable platform. Each processor in the pipelined MPSoC has a number of configurations resulting from customisable options such as custom instructions, cache configurations, etc. A design point of a pipelined MPSoC is then one of the combinations of these processor configurations. The goal is to select one configuration for each processor in the pipelined MPSoC to have the optimal combination of processor configurations (the optimal design point) for a given objective function such as minimum area or maximum throughput. The aim of this monograph is to optimise a pipelined MPSoC with such a design space for area footprint and energy consumption under performance constraints.[3]

This monograph proposes design-time and run-time optimisations targeted at different objective functions. At first, a pipelined MPSoC is optimised for area footprint under either an execution time, a latency or a throughput constraint by selection of the most suitable processor configurations during its design space exploration. Then, such a design-time optimised pipelined MPSoC is augmented with run-time adaptability to deactivate idle processors at run-time to reduce energy consumption. Here, the fact that not all the processors will be utilised at all times under a dynamic workload is exploited by the proposed run-time management techniques. Finally,

[3] Note that partitioning and mapping of a multimedia application on a pipelined MPSoC is done either manually or semi-automatically [99–103].

pipelined MPSoCs optimised for different multimedia applications are combined into a single multi-mode pipelined MPSoC for further reduction of area footprint. In particular, this monograph introduces:

1. **Design space exploration of pipelined MPSoCs.** For a pipelined MPSoC with 5 processors where each processor has 100 configurations, 10^{10} combinations of processor configurations are possible. To explore such a large design space, quick availability/evaluation of performance of design points and clever algorithms are required as full-system, cycle-accurate simulation and exhaustive search of all the design points is not feasible.

   - Analytical models are proposed to estimate execution time, latency and throughput of a pipelined MPSoC's design point using latencies of individual processor configurations, avoiding slow, full-system, cycle-accurate simulations of all the design points. For effective use of these analytical models, latencies of individual processor configurations should be available. Two estimation methods are proposed to gather latencies of processor configurations with minimal number of simulations. The first method simulates all the individual processor configurations once, while the second method simulates only a subset of processor configurations and then uses a processor analytical model to estimate the latencies of the processor configurations.
   - Three exploration techniques are proposed for optimisation of a pipelined MPSoC's area footprint. The first two techniques minimise area footprint under an execution time constraint and a latency constraint by exploiting execution time and latency analytical models in Integer Linear Programming (ILP) formulations. The third technique uses an algorithm to minimise area footprint under a throughput constraint by exploiting the throughput analytical model. Combined use of analytical models, estimation methods, and exploration algorithms enabled quick exploration of design spaces containing up to 10^{18} design points.

2. **Adaptive pipelined MPSoCs.** Area footprint optimised pipelined MPSoCs lack adaptability to dynamic workload of multimedia applications, and hence will keep all the processors active at all times, resulting in increased energy consumption.

   - An adaptive pipelined MPSoC architecture is proposed to enable run-time adaptability. Each stage with significant run-time variations in workload is implemented using *Main Processors* and *Auxiliary Processors*, where the main processor uses differing number of auxiliary processors considering run-time workload variations. Such an architecture allows the main processor of a stage to manage its auxiliary processors, independent of other stages, enabling the use of scalable, distributed run-time managers.
   - A run-time processor manager is proposed to predict the idle auxiliary processors of a main processor at run-time. The processor management heuristic uses a combination of the application's execution and knowledge (algorithmic and data properties), and information from off-line profiling and statistical analysis

to proactively predict the number of auxiliary processors that should be used. The idle auxiliary processors are either clock- or power-gated to reduce energy consumption.

- A run-time power manager (built on top of the processor manager) is proposed where auxiliary processors have multiple power states, trading-off overhead of the transition to power states with their possible energy reductions rather than just using clock-gating or power-gating. The power management heuristic forecasts at run-time, the idle duration of an idle auxiliary processor using the application's knowledge (algorithmic and data properties) so that the most suitable power state can be selected using the information from off-line analytical analysis of the power states. Experiments illustrated that adaptive pipelined MPSoC with processor manager saved up to 39% energy consumption compared to a pipelined MPSoC without run-time adaptability. Furthermore, use of the power manager (with the processor manager) reduced up to a further 40 % energy consumption compared to the use of only the processor manager.

3. **Multi-mode pipelined MPSoCs.** The area footprint and energy consumption optimisations targeted a single pipelined MPSoC executing one multimedia application. To further reduce area footprint, processors and FIFO buffers of multiple pipelined MPSoCs, designed for multiple multimedia applications, are shared when their use is mutually exclusive by creating a multi-mode pipelined MPSoC. Pipelined MPSoCs are represented as graphs to capture the number of processors, and number, size and connection of the FIFO buffers. Three heuristics are proposed to find maximal overlap between the graphs where two of them greedily find the overlap while the third one, based on maximum weight clique approach, finds an optimal overlap at the cost of higher running time. The results indicate significant area saving (up to 62 % processor area, 57 % FIFO area and 44 processor/FIFO ports) with minuscule degradation of system throughput (up to 2 %) and latency (up to 2 %), and an increase in energy consumption (up to 3 %) when compared to individual pipelined MPSoCs.

1.5 Monograph Outline

The remainder of the monograph is outlined as follows. Chapter 2 provides the necessary literature survey of notable homogeneous and heterogeneous MPSoCs. The literature survey also reports various design space exploration and run-time adaptability techniques for heterogeneous MPSoCs in general and pipelined MPSoCs in particular. Chapter 3 provides a philosophical overview of the research reported in this monograph.

Chapter 4 proposes analytical models for execution time, latency and throughput of a pipelined MPSoC. Chapter 4 also introduces two estimation methods to reduce the number of full-system cycle-accurate simulations of a pipelined MPSoC to aid quick design space exploration. Chapter 5 builds upon the analytical models and estimation methods by proposing techniques for area footprint minimisation of a pipelined MPSoC under performance constraints.

The adaptive pipelined MPSoC architecture is described in Chap. 6, in addition to a run-time processor manager and its heuristics. Chapter 7 describes the run-time power manager and its heuristics. These chapters also present a system-level overview and implementation of an adaptive pipelined MPSoC for an H.264 video encoder.

Chapter 8 presents the case for multi-mode pipelined MPSoC, followed by the heuristics for merging of individual pipelined MPSoCs. The final chapter, Chap. 9, summarises the research conducted during the course of this monograph. Chapter 9 also presents the author's proposals for future work.

1.6 Summary

This chapter introduced multimedia applications and their architectures currently in use in academia and industry. The challenges in design of heterogeneous MPSoCs for multimedia were discussed to motivate the need for selection of a multimedia platform and its optimisation for reduced area footprint and reduced energy consumption using design space exploration and run-time adaptability. Lastly, the chapter stated the contributions of the monograph.

References

1. M. Weiser, The computer for the 21st century. SIGMOBILE Mob. Comput. Commun. Rev. **3**, 3–11 (1999)
2. M. Morales, S. Rau, M.J. Palma, M. Venkatesan, F. Pulskamp, A. Dugar, Worldwide intelligent systems 2011–2015 forecast: the next big opportunity. Technical report, International Data Corporation, September 2011
3. C. Kozyrakis and D. Patterson, Vector vs. superscalar and vliw architectures for embedded multimedia benchmarks, in *MICRO 35: Proceedings of the 35th Annual ACM/IEEE International Symposium on Microarchitecture*, Los Alamitos, CA, USA, pp. 283–293 (IEEE Computer Society Press, 2002)
4. K. Karuri, R. Leupers, *Application analysis tools for ASIP design: application profiling and instruction-set customization* (Springer, New York, 2011)
5. K. Keutzer, S. Malik, A. Newton, From asic to asip: the next design discontinuity, in *Proceedings of the 2002 IEEE International Conference on Computer Design: VLSI in Computers and Processors*, pp. 84–90 (2002)
6. M. Shafique, Architectures for adaptive low-power embedded multimedia systems. Ph.D. thesis, Karlsruhe Institute of Technology, Germany (2011)
7. J. Meehan, S. Busch, J. Noel, F. Noraz, Multimedia ip architecture trends in the mobile multimedia consumer device. Image Commun. **25**, 317–324 (2010)
8. K. Willner, K. Ugur, M. Salmimaa, A. Hallapuro, J. Lainema, Mobile 3D video using mvc and n800 internet tablet, in *3DTV Conference: The True Vision—Capture, Transmission and Display of 3D Video, 2008*, pp. 69–72, May 2008
9. Nokia, Mobile 3d video (2012), http://research.nokia.com/

10. International Telecommunucation Union, Advanced video coding for generic audiovisual services, in *Recommendation H.264 and ISO/IEC 14496–10:2005* (2005)
11. Joint Video Team of ISO/IEC MPEG and I.-T. VCEG, Jvt-ab204: Joint draft 8.0 on multiview video coding (2008)
12. T. Wiegand, G. Sullivan, G. Bjontegaard, A. Luthra, Overview of the h.264/avc video coding standard. IEEE Trans. Circ. Syst. Video Technol. **13**, 560–576 (2003)
13. J. Ostermann, J. Bormans, P. List, D. Marpe, M. Narroschke, F. Pereira, T. Stockhammer, T. Wedi, Video coding with h.264/avc: tools, performance, and complexity. IEEE Circ. Syst. Mag. **4**, 7–28 (2004)
14. C. Van Berkel, Multi-core for mobile phones, in *Proceedings of the Design, Automation Test in Europe Conference Exhibition, 2009 (DATE '09)*, pp. 1260–1265, April 2009
15. Mentor Graphics, Modelsim, http://www.mentor.com
16. Synopsys, Design compiler, http://www.synopsys.com
17. Cadence, Virtuoso platform, http://www.cadence.com
18. T. von Sydow, B. Neumann, H. Blume, T. G. Noll, Quantitative analysis of embedded fpga-architectures for arithmetic, in *Proceedings of the International Conference on Application-Specific Systems, Architectures and Processors, 2006 (ASAP '06)*, pp. 125–131, September 2006
19. Y.-S. Huang, B.-C. Chieu, Architecture for video coding on a processor with an arm and dsp cores. Multimedia Tools Appl. **54**, 527–543 (2011)
20. Texas Instruments, Texas instruments dsps, http://www.ti.com
21. D. Talla, L. John, V. Lapinskii, B. Evans, Evaluating signal processing and multimedia applications on simd, vliw and superscalar architectures, in *Proceedings of the 2000 International Conference on Computer Design*, pp. 163–172 (2000)
22. Freescale, Freescale dsps, http://www.freescale.com
23. Analog Devices, Analog devices dsps, http://www.analog.com
24. NXP, Nxp trimedia architecture (2012), http://www.nxp.com
25. Tenslica, Connx vectra lx dsp engine, http://www.tensilica.com
26. Tensilica, Xtensa customizable processor, http://www.tensilica.com
27. U.J. Kapasi, S. Rixner, W.J. Dally, B. Khailany, J.H. Ahn, P. Mattson, J.D. Owens, Programmable stream processors. Computer **36**, 54–62 (2003)
28. G.G. Lee, Y.-K. Chen, M. Mattavelli, E.S. Jang, Algorithm/architecture co-exploration of visual computing on emergent platforms: overview and future prospects. IEEE Trans. Circ. Sys. Video Technol. **19**, 1576–1587 (2009)
29. G. R. Stewart, Implementing video compression algorithms on reconfigurable devices. Ph.D. thesis, University of Glassgow (2009)
30. S. Hu, Z. Zhang, M. Zhang, T. Sheng, Optimization of memory allocation for h.264 video decoder on digital signal processors, in *Proceedings of the Congress on Image and Signal Processing, 2008 (CISP '08)*, vol. 2, pp. 71–75, May 2008
31. P. Ienne, R. Leupers, *Customizable Embedded Processors: Design Technologies and Applications (Systems on Silicon)* (Morgan Kaufmann Publishers, San Mateo, 2006)
32. C. Valderrama, L. Jojczyk, P. Possa, J. Gazzano, Fpga and asic convergence, in *Proceedings of the 2011 7th Southern Conference on Programmable Logic (SPL)*, pp. 269–274, April 2011
33. Tensilica, Xtensa lx benchmarks, http://www.tensilica.com
34. S. Damaraju, V. George, S. Jahagirdar, T. Khondker, R. Milstrey, S. Sarkar, S. Siers, I. Stolero, A. Subbiah, A 22nm ia multi-cpu and gpu system-on-chip, in *Proceedings of the IEEE International Solid-State Circuits Conference Digest of Technical Papers (ISSCC)*, pp. 56–57, February 2012
35. J. Henkel, S. Parameswaran, *Designing Embedded Processors: A Low Power Perspective* (Springer, New York, 2007)
36. S. Saponara, L. Fanucci, S. Marsi, G. Ramponi, D. Kammler, E. Witte, Application-specific instruction-set processor for retinex-like image and video processing. IEEE Trans. Circ. Syst. II: Express Briefs **54**, 596–600 (2007)

37. S.D. Kim, M.H. Sunwoo, Asip approach for implementation of h.264/avc. J. Signal Process. Syst. **50**(1), 53–67 (2008)
38. J. Janhunen, O. Silven, M. Juntti, M. Myllyla, Software defined radio implementation of k-best list sphere detector algorithm, in *Proceedings of the International Conference on Embedded Computer Systems: Architectures, Modeling, and Simulation (SAMOS 2008)*, pp. 100–107, July 2008
39. A. Portero, G. Talavera, M. Moreno, J. Carrabina, F. Catthoor, Methodology for energy-flexibility space exploration and mapping of multimedia applications to single-processor platform styles. IEEE Trans. Circ. Syst. Video Technol. **21**, 1027–1039 (2011)
40. ARC, Arc configurable processors, http://www.arc.com
41. CoWare, Lisatek, http://www.coware.com/
42. Mips Technologies, Mips corextend processor, http://www.mips.com
43. Target Compiler Technologies, Ip designer, http://www.retarget.com
44. Asip Solutions, Asip meister, http://www.asip-solutions.com
45. University of California Irvine, Expression adl, http://www.ics.uci.edu/express/
46. Xilinx, Microblaze soft core, http://www.xilinx.com
47. Tensilica, XPRES Compiler, http://www.tensilica.com/
48. S. Vassiliadis, S. Wong, G. Gaydadjiev, K. Bertels, G. Kuzmanov, E. Panainte, The molen polymorphic processor. IEEE Trans. Comput. **53**, 1363–1375 (2004)
49. R. Lysecky, G. Stitt, F. Vahid, Warp processors. ACM Trans. Des. Autom. Electron. Syst. **11**, 659–681 (2006)
50. L. Bauer, M. Shafique, J. Henkel, Rispp: a run-time adaptive reconfigurable embedded processor, in *proceedings of the International Conference on Field Programmable Logic and Applications, 2009 (FPL 2009)*, pp. 725–726, 31 August–2 September 2009
51. Altera, Nios processor, http://www.altera.com
52. Microsoft Research, Emips: a dynamically extensible processor, http://research.microsoft.com/
53. Stretch, Reconfigurable processors, http://www.stretchinc.com
54. H. Amano, A survey on dynamically reconfigurable processors, in *Proceedings of the IEICE Transactions* (2006)
55. H.P. Huynh, T. Mitra, Runtime adaptive extensible embedded processors: a survey, in *Proceedings of the 9th International Workshop on Embedded Computer Systems: Architectures, Modeling, and Simulation (SAMOS '09)*, pp. 215–225 (Springer, Berlin, 2009)
56. H. Sutter, The free lunch is over, http://www.gotw.ca/publications/concurrency-ddj.htm
57. S.L. Shee, A. Erdos, S. Parameswaran, Heterogeneous multiprocessor implementations for jpeg: a case study, in *CODES+ISSS '06: Proceedings of the 4th International Conference on Hardware/Software Codesign and System Synthesis*, pp. 217–222 (ACM, New York, 2006)
58. S.L. Shee, A. Erdos, S. Parameswaran, Architectural exploration of heterogeneous multiprocessor systems for jpeg. Int. J. Parallel Prog. **36**(1), 140–162 (2008)
59. H.C. Doan, H. Javaid, S. Parameswaran, Multi-asip based parallel and scalable implementation of motion estimation kernel for high definition videos, in *Proceedings of the 2011 9th IEEE Symposium on Embedded Systems for Real-Time Multimedia (ESTIMedia)*, pp. 56–65, October 2011
60. F. J. Pollack, New microarchitecture challenges in the coming generations of cmos process technologies (keynote address) (abstract only), in *Proceedings of the 32nd Annual ACM/IEEE International Symposium on Microarchitecture (MICRO 32)*, p. 2 (IEEE Computer Society, Washington DC, 1999)
61. W. Knight, Two heads are better than one (dual-core processors). IEE Rev. **51**, 32–35 (2005)
62. P. Gepner, D. Fraser, M. Kowalik, R. Tylman, New multi-core intel xeon processors help design energy efficient solution for high performance computing, in *Proceedings of the International Multiconference on Computer Science and Information Technology, 2009 (IMCSIT '09)*, pp. 567–571, October 2009
63. S. Borkar, Thousand core chips: a technology perspective, in *Proceedings of the 44th Annual Design Automation Conference (DAC '07)*, pp. 746–749 (ACM, New York, 2007)

64. K. Asanovic, R. Bodik, J. Demmel, T. Keaveny, K. Keutzer, J. Kubiatowicz, N. Morgan, D. Patterson, K. Sen, J. Wawrzynek, D. Wessel, K. Yelick, A view of the parallel computing landscape. Commun. ACM **52**, 56–67 (2009)
65. G. Martin, Multi-processor soc-based design methodologies using configurable and extensible processors. J. Signal Process. Syst. **53**(1–2), 113–127 (2008)
66. Y.-K. Chen, C. Chakrabarti, S. Bhattacharyya, B. Bougard, Signal processing on platforms with multiple cores, part 1: overview and methodologies (from the guest editors). IEEE Signal Process. Mag. **26**, 24–25 (2009)
67. International Technology Roadmap for Semiconductors, System drivers (2011), http://www.itrs.net
68. J. Goodacre, A. Sloss, Parallelism and the arm instruction set architecture. Computer **38**, 42–50 (2005)
69. U. Kapasi, W. Dally, S. Rixner, J. Owens, B. Khailany, The imagine stream processor, in *Proceedings of the 2002 IEEE International Conference on Computer Design: VLSI in Computers and Processors*, pp. 282–288 (2002)
70. Tilera, Tilepro64 multicore processor product brief, http://www.tilera.com
71. Intel, Single-chip cloud computer, http://www.intel.com
72. W. Wolf, A. Jerraya, G. Martin, Multiprocessor system-on-chip (mpsoc) technology. IEEE Trans. Comput. Aided Des. Integr. Circ. Syst. **27**, 1701–1713 (2008)
73. R. Kumar, D.M. Tullsen, P. Ranganathan, N.P. Jouppi, K.I. Farkas, Single-isa heterogeneous multi-core architectures for multithreaded workload performance. SIGARCH Comput. Archit. News **32**, 64 (2004)
74. R. Kumar, D. M. Tullsen, N. P. Jouppi, Core architecture optimization for heterogeneous chip multiprocessors, in *Proceedings of the 15th International Conference on Parallel Architectures and Compilation Techniques (PACT '06)*, pp. 23–32 (ACM, New York, 2006)
75. M. Hill, M. Marty, Amdahl's law in the multicore era. Computer **41**, 33–38 (2008)
76. F. Sun, S. Ravi, A. Raghunathan, N.K. Jha, Synthesis of application-specific heterogeneous multiprocessor architectures using extensible processors, in *VLSID '05: Proceedings of the 18th International Conference on VLSI Design Held Jointly with 4th International Conference on Embedded Systems Design*, pp. 551–556 (IEEE Computer Society, Washington DC, 2005)
77. R. Hameed, W. Qadeer, M. Wachs, O. Azizi, A. Solomatnikov, B.C. Lee, S. Richardson, C. Kozyrakis, M. Horowitz, Understanding sources of ineffciency in general-purpose chips. Commun. ACM **54**, 85–93 (2011)
78. H. P. Hofstee, Power efficient processor architecture and the cell processor, in *Proceedings of the 11th International Symposium on High-Performance Computer Architecture*, pp. 258–262 (IEEE Computer Society, 2005)
79. Intel, Ixp network processors, http://www.intel.com
80. Nxp Semiconductors, Nexperia media processor, http://www.nxp.com
81. Texas Instruments, Omap mobile processors, http://www.ti.com/
82. STMicroelectronics, Nomadik application processor, http://www.st.com
83. P. Flake, S. Davidmann, F. Schirrmeister, System-level exploration tools for mpsoc designs, in *DAC '06: Proceedings of the 43rd annual Design Automation Conference*, pp. 286–287, (ACM, New York, 2006)
84. G. Kahn, The semantics of a simple language for parallel programming, in *Proceedings of the IFIP Congress on Information Processing '74*, pp. 471–475 (1974)
85. E.A. Lee, D.G. Messerschmitt, Synchronous data flow. Proc. IEEE **75**(9), 1235–1245 (1987)
86. W. Thies, M. Karczmarek, S.P. Amarasinghe, Streamit: a language for streaming applications, in *Proceedings of the 11th International Conference on Compiler Construction, CC '02*, pp. 179–196 (Springer, 2002)
87. A. Sangiovanni-Vincentelli, Quo vadis, sld? reasoning about the trends and challenges of system level design. Proc. IEEE **95**, 467–506 (2007)
88. H. Guo, S. Parameswaran, Balancing system level pipelines with stage voltage scaling, in *Proceedings of the IEEE Computer Society Annual Symposium on VLSI: New Frontiers in VLSI Design, ISVLSI '05* (2005)

89. S. Carta, A. Alimonda, A. Pisano, A. Acquaviva, L. Benini, A control theoretic approach to energy-efficient pipelined computation in mpsocs, ACM Trans. Embedded Comput. Syst. Article id 27: **6**(4), 28 (2007)
90. A. Alimonda, S. Carta, A. Acquaviva, A. Pisano, L. Benini, A feedback-based approach to dvfs in data-flow applications. IEEE Trans. CAD Integr. Circ. Syst. **28**(11), 1691–1704 (2009)
91. S. L. Shee, S. Parameswaran, Design methodology for pipelined heterogeneous multiprocessor system, in *DAC '07: Proceedings of the 44th Annual Conference on Design Automation*, pp. 811–816 (2007)
92. H. Javaid, A. Ignjatovic, S. Parameswaran, Rapid design space exploration of application specific heterogeneous pipelined multiprocessor systems. Trans. Comput. Aided Des. Integr. Circ. Syst. **29**, 1777–1789 (2010)
93. I. Karkowski, H. Corporaal, Design of heterogenous multi-processor embedded systems: applying functional pipelining, in *PACT '97: Proceedings of the 1997 International Conference on Parallel Architectures and Compilation Techniques* (IEEE Computer Society, 1997)
94. M.I. Gordon, W. Thies, S. Amarasinghe, Exploiting coarse-grained task, data, and pipeline parallelism in stream programs. SIGPLAN Not. **41**, 151–162 (2006)
95. H. Park, Y. Park, S. Mahlke, Polymorphic pipeline array: a flexible multicore accelerator with virtualized execution for mobile multimedia applications, in *Proceedings of the 42nd Annual IEEE/ACM International Symposium on Microarchitecture (MICRO 42)*, pp. 370–380 (ACM, New York, 2009)
96. M.A. Suleman, M.K. Qureshi, Khubaib, Y.N. Patt, Feedback-directed pipeline parallelism, in *Proceedings of the 19th International Conference on Parallel Architectures and Compilation Techniques (PACT '10)*, pp. 147–156 (ACM, New York, 2010)
97. 4g applications, architectures, design methodology and tools for mpsoc, in *DATE '06: Proceedings of the Conference on Design, Automation and Test in Europe*, pp. 830–831 (European Design and Automation Association, Leuven, 2006)
98. G. Goossens, Multi-asip socs: or how to design ultra-low power architectures for wireless and multi-media systems, in *Proceedings of the 2007 International Symposium on System-on-Chip*, p. 1, November 2007
99. S. Verdoolaege, H. Nikolov, T. Stefanov, Pn: a tool for improved derivation of process networks. EURASIP J. Embedded Syst. **2007**, 19 (2007)
100. D. Cordes, A. Heinig, P. Marwedel, A. Mallik, Automatic extraction of pipeline parallelism for embedded software using linear programming, in *Proceedings of the 2011 IEEE 17th International Conference on Parallel and Distributed Systems (ICPADS)*, pp. 699–706, December 2011
101. M. Kudlur, S. Mahlke, Orchestrating the execution of stream programs on multicore platforms, in *Proceedings of the ACM SIGPLAN Conference on Programming Language Design and Implementation (PLDI '08)* (2008)
102. M. Hashemi, S. Ghiasi, Throughput-driven synthesis of embedded software for pipelined execution on multicore architectures. ACM Trans. Embed. Comput. Syst. **8**, 11:1–11:35 (2009)
103. S. M. Farhad, Y. Ko, B. Burgstaller, B. Scholz, Orchestration by approximation: mapping stream programs onto multicore architectures, in *Proceedings of the 16th International Conference on Architectural Support for Programming Languages and Operating Systems (ASPLOS '11)* (2011)

Chapter 2
Literature Survey

Many researchers have looked at design space exploration and run-time adaptability of MPSoCs. This chapter provides the necessary literature survey, starting with homogeneous and heterogeneous MPSoCs. Focus is then directed to design space exploration techniques such as linear programming and heuristics. The chapter concludes with run-time resource and power management techniques for MPSoCs.

2.1 Homogeneous MPSoCs

Homogeneous MPSoCs, also referred to as chip multiprocessors, use the paradigm of Symmetric MultiProcessing (SMP) and employ identical processors with same Instruction Set Architecture (ISA). Multimedia applications, such as JPEG, MP3, and H.264, contain common sub-kernels like Fast Fourier Transform (FFT) and Discrete Cosine Transform (DCT). Therefore, researchers have exploited these commonalities to include support for such sub-kernels in homogeneous MPSoCs for performance/energy efficient implementation of multimedia applications.

Stanford's Imagine [1, 2] is a programmable stream processor with 48 Arithmetic Logic Units (ALUs) consisting of floating-point adders, multipliers and divide square-root units. These ALUs are arranged into eight clusters which are interfaced with a local register file and a stream register file to provide the memory bandwidth required of multimedia applications. The eight clusters work in a SIMD manner with six-way VLIW instructions per cluster. Imagine has been illustrated to achieve 32 GOPS and 16 GOPS for single precision (matrix multiplication, etc.) and 16-bit fixed-point (2D DCT, etc.) applications respectively, running at a frequency of 400 MHz. Although Imagine is a stream processor, it can be classified as a homogeneous MPSoC due to the replication of identical clusters. Imagine is programmed in a stream model where applications are represented as a set of sub-kernels that consume and produce data streams. KernelC and StreamC programming languages were developed for easy programming of Imagine.

H. Javaid and S. Parameswaran, *Pipelined Multiprocessor System-on-Chip for Multimedia*, DOI: 10.1007/978-3-319-01113-4_2,

The RAW [3] processor from MIT is another example of a homogeneous MPSoC. It contains sixteen identical, programmable tiles where each tile has an in-order, single issue, eight stage pipeline, MIPS-like processor, local data and instruction caches, and static and dynamic routers. These tiles are arranged in a 4×4 grid and are connected through a Network-on-Chip (NoC) which is designed to run at high frequencies and to be scalable to even a thousand tiles. RAW has been shown to achieve from a $2\times$ to a $100\times$ performance improvement depending on the amount of data- and instruction-, and pipeline-level (stream-level) parallelisms [4]. Like Imagine, a compiler was developed to exploit these parallelisms. Later on, a backend for StreamIt [5] (a stream programming language) compiler was developed for RAW [4, 6].

TILEPro64 [7], much like RAW, features an 8×8 grid with 64 tiles where each tile contains a three-way VLIW processor supporting SIMD instructions on 32-, 16- and 8-bit data. In addition, each tile has on-chip separate L1 instruction and data caches, unified L2 cache, and a switch that connects the tile to a power-efficient mesh network. Idle tiles can be put into a low-power state to reduce energy consumption. TILEPro64 also features Tilera's dynamic distributed cache technology which provides a $2\times$ improvement in cache coherence performance over traditional coherence protocols. Software tools are provided for application analysis and compilation, although a designer has to manually partition the application. TILEPro has been shown to achieve more than 400 BOPS which translates to 15 Gbps of SNORT processing, 20 Gbps of nProbe processing and H.264 encoding of 10 HD (1080p) video streams at 30 fps [7, 8]. Several variants of the architecture are available from Tilera where the number of tiles range from 16 to 100.

Intel recently developed a research prototype of a tiled, homogeneous MPSoC, which was named Single-chip Cloud Computer (SCC) [9]. The SCC consists of 24 tiles organised into a 4×6 grid. A tile contains a pair of Pentium processors, each with independent, separate L1 instruction and data caches and a unified L2 cache. In addition, each tile contains a message passing buffer and a router for efficient inter-processor communication, but without the support for built-in (hardware) cache coherency. The most notable feature of SCC is the availability of multiple voltage and frequency domains for implementation of Dynamic Voltage and Frequency Scaling (DVFS) technique for low power consumption. SCC is still a research platform and several researches have being conducted recently to study performance and power management [10, 11], programming models [12, 13], and other related issues in large MPSoCs (many-core platforms).

Since homogeneous MPSoCs replicate identical tiles, they are scalable and easy to program. However, they also have a larger area footprint and higher power consumption because no single type of tile or processor can be well suited to every multimedia application [14]. Therefore, heterogeneous MPSoCs exploit customisation of processing elements and are more suitable for embedded devices because of stringent area footprint and power consumption constraints [15].

2.2 Heterogeneous MPSoCs

Numerous heterogeneous MPSoCs have been proposed because the domain of multimedia consists of a diverse set of applications. For example, the architectural requirements of networking applications are considerably different from those of audio/video applications. This section provides a brief overview of some of the popular heterogeneous MPSoCs designed for multimedia.

The C-5 network processor [16] is one of the early examples of heterogeneous MPSoCs, designed for packet processing applications in networking. It consists of an executive processor, a fabric processor and sixteen channel processors, with additional specialised buffer management, queue management and table lookup units. The executive processor (a RISC based GPP) serves as the central system manager and provides the standard system interfaces. The fabric processor extends the interfacing capability of C-5 by providing a high-speed network interface port. The main crux of C-5 lies in the channel processors, grouped into clusters of four, that receive, process and transmit network data. The specialised units perform management tasks so that the processors can work together efficiently. The C-5 network processor can process up to 15 million packets per second [16].

Intel's IXP [17] network processor contains Intel's XScale processor for general-purpose processing and two Network Processing Elements (NPEs) for packet processing. Each NPE contains eight micro-engines that are specialised functional units and act as hardware accelerators. The NPE is equipped with separate instruction and data memories, and hardware based multi-threading for high performance. In addition, IXP contains hardware accelerators for popular cryptography algorithms to provide security features. Cisco's Silicon Packet Processor (SPP) [18] and QuantumFlow processor [19] used in its CRS-1 and CRS-2 routers are other examples of heterogeneous MPSoCs. SPP and QuantumFlow employ 188 and 40 multi-threaded Xtensa LX processors [20], which are extremely customised for network data processing.

The Viper [21] from Philips is an early example of a video MPSoC. It consists of two main processors (a MIPS based GPP and a Timedia VLIW DSP [22]), various audio/video hardware blocks (accelerators) and a number of standard interfaces. The GPP runs an operating system and acts as a master to Trimedia DSP and audio/video hardware accelerators to process HD resolution videos. Later, a configurable heterogeneous MPSoC platform named Eclipse was proposed by Philips where the number and types of domain-specific hardware blocks, bus widths, memory sizes and similar parameters can be set by a designer. The aim was to provide a platform that can be tuned to an application without major architectural redesign effort. Eclipse contains a GPP-DSP (RISC-VLIW) based main processor that acts as the master to a number of weakly programmable, multitasking, domain-specific hardware blocks, referred to as coprocessors. For example, an instance of Eclipse may use Discrete Cosine Transform (DCT) and Motion Estimation (ME) coprocessors only for an MPEG decoder. Eclipse uses the Kahn Processor Network (KPN) as the application model and provides task scheduling and communication synchronisation methods for efficient utilisation of the coprocessors. Eclipse, when configured for video decoding, consumed less than 240 mW for simultaneous real-time decoding of two HD MPEG streams. NXP Semiconductor's Nexperia [23] architecture is based on Viper and Eclipse.

STMicroelectronic's Nomadik [24] is another heterogeneous MPSoC that was designed for mobile phones. It contains a master processor (an ARM based GPP) in addition to two slave DSPs: one for audio and the other for video. These DSPs act as hardware accelerators for audio and video applications. The video DSP itself is a heterogeneous MPSoC, consisting of a DSP and several hardware accelerators like an image pre-/post-processor and a video encoder/decoder. The audio DSP uses a single DSP because audio applications require relatively smaller amounts of computational power. The Nomadik processor has been succeeded by the NovaThor platform [25] which contains an ARM Cortex-A9 MPCore processor [26], two DSPs and an ARM Mali GPU [27] with NEON SIMD engine [28] for low-power, flexible multimedia processing up to HD resolution.

The CELL [29] is a heterogeneous MPSoC with two types of processing elements. The Power Processor Element (PPE) runs an operating system to provide services such as memory management and thread scheduling. The Synergistic Processor Elements (SPEs), on the other hand, are extremely specialised units that are based on VLIW and SIMD concepts and function as hardware accelerators. The CELL processor can exploit instruction-, data-, task- and memory-level parallelisms with a combination of a PPE, SPEs and Direct Memory Access (DMA) engines to provide a high performance, low-power implementation platform for multimedia applications.

NVIDIA's Tegra [30] is an example of one of the recent heterogeneous MPSoCs for multimedia in mobile phones. It consists of seven types of processing units for audio, video, image, graphics and general-purpose processing. More specifically, Tegra includes an ARM Cortex-A9 MPCore for general-purpose applications, an ARM7 processor for computationally less intensive and system management tasks, an ultra low-power Graphics Processing Unit (GPU), an audio processor, an Image Signal Processor (ISP), an HD video decoding processor and an HD video encoding processor. Tegra also includes a system-level power management module that shuts down idle processing units. Inclusion of such a diverse set of processing units and the power management unit allows Tegra to deliver high performance with ultra low power consumption. OMAP [31] is another recent heterogeneous MPSoC platform from Texas Instrument. Like Tegra, it combines several types of specialised processing units for specific tasks. It contains a Cortex-A9 MPCore, an Image Video Audio (IVA) hardware accelerator, an ISP, a PowerVR processor [32] based graphics accelerator, and a display sub-system. The IVA itself is a heterogeneous MPSoC with a power-optimised, multi-mode hardware accelerator and a DSP processor. OMAP is also equipped with Texas Instrument's SmartReflex power and performance management technology which includes adaptive techniques for run-time control of frequency, voltage and power to deliver required performance with ultra low power consumption.

There have also been non-commercial efforts in the field of heterogeneous MPSoCs for multimedia. Strik et al. [33] proposed a real-time video and graphics system composed of a control processing sub-system (host), signal processing sub-system and a memory sub-system. They used a reconfigurable NoC in the MPSoC to enable concurrent processing of 25 video streams in real-time. The heterogeneous MPSoC for HDTV proposed in [34] uses five processors communicating through FIFO buffers and shared memories. The proposed MPSoC was shown to deliver the required performance through a combination of application partitioning, customised VLIW DSPs and a custom two level scratchpad based memory hierarchy.

Wu et al. [35] proposed MediaDSP, a scalable architecture consisting of two types of processors, domain-specific hardware accelerators, a banked memory hierarchy, an on-chip crossbar network and DMA engines. The authors demonstrated research prototypes of a single-issue DSP with an ALU and a MAC unit, and a microcode based dual-issue four-way SIMD DSP. The scalability of MediaDSP comes from the on-chip network that can organise the processors in various topologies such as a pipeline configuration or a master-slave configuration. MediaDSP exploits data-, instruction-, memory- and task-level parallelisms and targets audio, video, gaming, user interface and computer vision applications.

Tumeo et al. [36] proposed a master-slave heterogeneous MPSoC consisting of two PowerPC processors [37], four Microblaze processors [38] and DMA engines. Each processor is connected to a local memory in addition to a shared memory. The processors synchronise with each other using interrupts. A software layer containing a microkernel is executed on the master processor to enable pipelined execution of multimedia applications. In addition, the microkernel is responsible for the transfer of data between the processors. The heterogeneity comes from different configurations of the processors which are selected by the designer according to the application to be executed on the MPSoC. Another master-slave architecture is the ePUMA [39] where the master is a GPP with a DMA engine, and is connected to eight SIMD DSP slaves with a ring bus in a star topology.

Most of the heterogeneous MPSoCs mentioned so far use master-slave configuration where a GPP is used as the master and a mix of ASICs, ASIPs and DSPs is used as slaves. Several researches have explored the pipeline configuration where processing elements are connected in a pipeline (chain). Park et al. [40] proposed a Polymorphic Pipeline Array (PPA) as an accelerator for multimedia applications, inspired from Coarse Grained Reconfigurable Arrays (CGRAs), but with both static and dynamic configurability. The PPA consists of an array of identical Processing Elements (PEs) that are tightly connected using a mesh-style interconnect, and a shared memory. A processor in PPA is made up of four PEs where the processor has an Instruction Set Architecture (ISA) and executes its own instruction stream. The heterogeneity is added by coalescing processors to create larger logical processors at run-time with the support of virtualised execution. Experiments with three multimedia applications showed that PPA can deliver required performance; however, it was less energy efficient (performance/power) than an ASIP (Tensilica's Diamond processor [41]).

Shee et al. [42, 43] performed a detailed comparison of master-slave configuration with pipeline configuration (will be referred to as pipelined MPSoCs for the rest of the monograph) through a case study on a JPEG encoder, which is a typical multimedia application. The empirical data clearly suggested that pipelined MPSoCs are more suitable for multimedia applications as they provided up to a 2 × performance improvement over master-slave MPSoCs. Shee et al. further illustrated that balancing workload across stages of the pipelined MPSoC with the use of ASIPs (that is, adding heterogeneity through customisation of the processors) can result in a 4.7 × performance improvement with a 3.1 × area increase compared to a 3.8 × performance improvement and a 7 × area increase of a homogeneous pipelined MPSoC. A recent paper by Hameed et al. [44] further analysed pipelined MPSoCs through a case study on an H.264 encoder for HD720p video resolution. They illustrated that extreme customisation of ASIPs can match performance of the pipelined MPSoC to that of an ASIC, but with 3 × energy consumption, which is the cost of reduced time-to-design, reduced time-to-market, flexibility and programmability of the pipelined MPSoC.

Pipelined MPSoCs exploit both data- and instruction-level parallelisms, which are abundant in multimedia applications, by using ASIPs with SIMD and VLIW techniques. More importantly, pipelined MPSoCs not only exploit task-level parallelism with the use of multiple ASIPs, but also pipeline-level (stream-level) parallelism of multimedia applications by arranging those ASIPs in stages of a pipeline. Therefore, pipelined MPSoCs have emerged as a viable implementation platform for multimedia applications [45–49]. Note that these pipelined MPSoCs can be used as standalone multimedia systems or as multimedia accelerators in commercial platforms. For example, a chip may contain multiple pipelined MPSoCs for video encoders and decoders; or OMAP, Tegra and other similar platforms may use pipelined MPSoCs to implement video encoder and decoder accelerators. Typically, pipelined MPSoCs will be used as multimedia accelerators because they are customised for specific multimedia applications.

2.3 Design Space Exploration

It is obvious that design and optimisation of heterogeneous MPSoCs is difficult due to the availability of a multitude of options such as application partitioning, MPSoC architecture, processor types and memory hierarchy. Therefore, there is a need for well-structured, systematic approaches to explore the design space resulting from these options. Often design space exploration is performed with multiple objectives and constraints, and the aim of quickly finding one or multiple (near-) optimal design points. This section provides an overview of typical design space exploration techniques used for heterogeneous MPSoCs, including pipelined MPSoCs.

2.3.1 Exact Approaches

Exact approaches in design space exploration search for the optimal design point and are typically based on Linear Programming (LP) [50]. In LP, variables that can take binary (0 or 1), integer or real values are used to represent parameters of the design space. The objective function is specified as a linear function of those variables while the constraints are described as a set of linear equalities and/or inequalities in the variables. These equations are then solved to find the values of the variables which are interpreted to get the final design point.

Batista et al. [51] formulated the problem of mapping and scheduling tasks of an application, which is represented as a task graph, onto a heterogeneous MPSoC with task-specific processors connected through a shared bus as a Mixed Integer Linear Programming (MILP) problem. The MILP model allowed pipelined execution of tasks, with minimisation of initiation interval (period), latency and/or MPSoC hardware cost as objective functions. A technique was also proposed to calculate upper and lower bounds on initiation interval to prune the design space which improved the MILP solver's time to search the optimal mapping and schedule of the tasks. Schwiegershausen et al. [52] also used MILP to explore the design space of a heterogeneous MPSoC with domain-specific blocks and processors. However, their objective function comprised of a weighted sum of period, latency and MPSoC area footprint (processors and busses) to prioritise the optimisation process based on the individual weights. The proposed MILP was tested by mapping H.261 video encoder on a heterogeneous MPSoC with four types of processors.

The problem of mapping the Kahn Process Network (KPN) representation of an application onto a heterogeneous MPSoC was formulated as a MILP problem in [53]. Their heterogeneous MPSoC used ASICs, GPPs or DSPs as compute units, and single and multiple buses, and crossbar switches as interconnection components. The aim of the MILP model was to map processes in KPN to suitable compute units and channels between processes to either local or shared memories, and selection of interconnection components to minimise MPSoC's hardware cost (cost of compute units, memories and interconnection components) under performance and bandwidth constraints. Kuang et al. [54] also targeted a heterogeneous MPSoC with ASICs, GPPs and DSPs, and a communication network; however, their aim was to simultaneously map and schedule application tasks to allow pipelined execution on the MPSoC. The authors proposed an ILP model to minimise MPSoC's area footprint under a throughput constraint.

Suhendra et al. [55] studied the problem of task mapping and scheduling in a heterogeneous MPSoC where each processor had a local scratchpad memory. Scratchpad memories are typically deployed in embedded devices due to their smaller area footprint, lower energy consumption and better timing predictability over caches [56]. An ILP formulation was proposed with the objective function of maximising application performance under an area constraint for the scratchpad memories. The ILP modelled mapping, scheduling and pipelined execution of tasks, and sizes of scratchpad memories and allocation of variables to them. Experiments with several multimedia applications showed that simultaneous optimisation of scratchpad memory sizes and task mapping/scheduling can improve performance by up to 80 % compared to the case when these optimisations are done separately.

Several works have also looked at the transformation of the application task graph during the mapping and scheduling problem. Ostler et al. [57] targeted mapping of application task graphs onto the Intel's IXP network processor [17] where each processing element is multi-threaded and has access to scratchpad, local and shared memories. Therefore, during the mapping problem, they considered merging and replication of tasks in addition to allocation of data to one of the memory types. An ILP formulation was proposed with the objective function of maximising the application throughput. Experiments with networking applications showed that task graph transformations coupled with data mapping can result in up to an $8\times$ performance improvement. The work in [58] considered a similar problem with some extensions and proposed an ILP formulation. They considered pipelined scheduling of tasks, and multi-bank register files rather than scratchpad memories as one of the memory types. Furthermore, their objective function was to minimise a weighted sum of throughput and latency because both of these are important for real-time multimedia applications. Yang et al. [59] also used ILP to formulate the problem of task mapping and scheduling with the consideration of data-level parallelism where appropriate tasks were duplicated. Their objective was to minimise the number of processors under task deadlines.

Reliability has become an important concern in design of MPSoCs. Tosun et al. [60] studied task mapping and scheduling considering fallibility (opposite to reliability) of the underlying heterogeneous MPSoC. They considered several objective functions and constraints such as maximising performance under energy consumption and fallibility constraints, minimising energy consumption under performance and fallibility constraints, and minimising fallibility under performance and energy consumption constraints. Dynamic Voltage Scaling (DVS) and task duplication were used to reduce energy consumption and fallibility of an application. A more comprehensive exploration approach, based on MILP, was proposed in [61] where the authors considered task mapping and pipelined scheduling, heterogeneous processing elements (differing error rates, clock frequencies and power consumptions), communication overheads and mutual exclusion from locks/critical sections. Their objective was to minimise energy consumption under performance and reliability constraints.

In contrast to the above mentioned works, the authors of [62] formulated the problem of mapping and pipelined scheduling of tasks on a message-oriented, distributed memory, shared bus heterogeneous MPSoC with special consideration to communication costs as an ILP problem and a Constraint Programming (CP) problem respectively. They showed that solving mapping and scheduling problems as a pure ILP or as a pure CP is much slower than the combined use of ILP and CP.

LP has also been used for the exploration of FPGA based (soft) MPSoCs. Wu et al. [63] proposed an ILP formulation to obtain the MPSoC architecture with mapping and scheduling of the tasks. In their work, the MPSoC design space consisted of heterogenous processors, memory configurations, point-to-point FIFO buffers, private busses and shared busses. The ILP model aimed for minimisation of FPGA area footprint under a performance constraint. This work was extended in [64] to consider instruction and data memory sizes as well because Block RAM (BRAM) in FPGAs is a limited resource that needs to be used judiciously. Furthermore, the

ILP model in [64] allows multiple objective functions targeting FPGA area footprint and memory size, and application execution time.

Parallel compilation of applications on MPSoCs has also been achieved using LP. An Integer Linear Programming (ILP) based compilation approach to parallelise loops in array-intensive applications on a shared memory, shared bus, homogeneous MPSoC was proposed in [65]. Their ILP model searches for the number of processors required to execute a loop with objective functions and constraints involving execution time and/or energy consumption. Their ILP also models the overhead involved in activating and deactivating processors at run-time as the number of processors changes from one loop to another during the execution of the application. Choi et al. [66] used ILP for compilation of streaming (multimedia) applications, which is represented as a graph, on a heterogeneous MPSoC with master-slave configuration. The ILP model was aimed at mapping and scheduling of tasks on processors, allocation of variables to local memory and generation of DMA transfers between processors with the objective function of minimising initiation interval (period) under memory and timing constraints. They also proposed an ILP formulation to minimise the number of processors under memory and timing constraints. Since they considered a large design space, several heuristics were used with ILP to keep exploration tractable and to obtain near-optimal design point quickly. The proposed compilation approach was implemented in SUIF compiler framework [67] and was tested with the compilation of Software Defined Radio (SDR) application on the CELL [29] MPSoC.

Design space exploration of pipelined MPSoCs have also been performed using ILP. Jin et al. [68] explored mapping of a multimedia application's task graph on an FPGA based homogeneous MPSoC using an ILP formulation. They used buses and point-to-point FIFO buffers for communication between the processors. The aim of the ILP model was to maximise the throughput of the application with a fixed number of processors. This work was improved by Cong et al. [69] by exploring not only the mapping but also partitioning of the task graph to find the minimum number of processors rather than fixing the number of processors in the MPSoC. Cong et al. developed several exact graph algorithms (without any variants of LP) based on labelling, clustering and packing techniques to minimise latency and the number of processors under a throughput constraint. Since multimedia applications exhibit run-time variations in execution time due to data-dependent behaviour, their application model associated probabilities with execution times of the tasks.

2.3.2 Heuristic Approaches

Linear programming based approaches are exhaustive in the worst case and can be slow for complex heterogeneous MPSoCs. Therefore, researchers have developed several heuristic approaches so as to rapidly and efficiently explore the design space at hand. Heuristic approaches do not guarantee an optimal solution; however, use of cleverly designed algorithms can provide remarkable improvements in exploration time with near-optimal solutions.

Banerjee et al. [70] considered mapping and pipelined scheduling of a multimedia application, which is represented as a Directed Acyclic Graph (DAG), onto a heterogeneous MPSoC with ASICs and DSPs. A two level hierarchal heuristic approach is proposed where a coarse-grained solution, in the number of pipeline stages in the MPSoC, is obtained by partitioning the DAG using the ratio cut partitioning technique. This initial solution is then refined through successive application of either architecture based partitioning or repartitioning, and retiming techniques. The heuristic is terminated once no more throughput improvement is observed. Experiments with typical multimedia applications illustrated that pipelined scheduling using heterogeneous processing elements improved throughput by several times over homogeneous processing elements.

Bakshi et al. [71, 72] also considered partitioning and pipelined scheduling of a task graph onto a heterogeneous MPSoC. Their heuristic approach maps all the tasks on processors first and then moves those that violate the throughput constraint to ASICs to obtain an initial mapping. Then, pipelined scheduling is performed to determine the number of pipeline stages based on a list scheduling algorithm. Unlike [71, 72], Jeon et al. [73] proposed to perform pipelining before partitioning and mapping of tasks on the heterogeneous MPSoC. In addition, they considered hardware sharing during mapping, and proposed an iterative heuristic approach to successively find more parallelism, while reducing area footprint through hardware sharing under a performance constraint.

The work in [74] considered a pipelined scheduling problem similar to [70], but with task duplication. The number of processors in the MPSoC was fixed beforehand and their objective function was to minimise latency rather than the throughput. A five step heuristic approach was proposed where the first three steps obtain an initial number of pipeline stages by clustering the application DAG. The fourth step then duplicates tasks to reduce latency if some of the processors are still free. However, if the number of clusters is more than the number of processors in the MPSoC, then cluster compaction is performed. The final step produces the pipelined schedule of the DAG.

Benoit et al. [75–79] proposed several heuristics for mapping and pipelined scheduling of linear application DAGs (containing only a single path) onto a heterogeneous MPSoC with a fixed number of homogeneous processors. The heterogeneity in the MPSoC was manifested by differing frequencies of processors and bandwidths of interprocessor links. The mapping and scheduling problem was formulated as an interval mapping problem where intervals, consisting of consecutive tasks, were scheduled on the processors to allow pipelined execution. Three heuristics were proposed to maximise the throughput [75] while a dynamic programming based algorithm was proposed to minimise the latency [76]. Several other heuristics were also proposed to minimise throughput under a latency constraint and to minimise latency under a throughput constraint, allowing bi-objective optimisations to find Pareto-optimal mappings and schedules of linear DAGs [77, 78]. The heuristics consisted of two steps: firstly, assign all the tasks to the fastest processor; and secondly, greedily split the largest interval to improve throughput or latency. The heuristics iteratively applied the second step and differed in the function chosen to split the interval. Benoit et al. [79] also proposed an ILP formulation to find optimal mapping and schedule for evaluation of the solutions found by the proposed heuristics.

Ko et al. [80] studied pipelined scheduling of multimedia applications with the focus on exploration of various pipeline configurations under latency and throughput constraints. They proposed Pipeline Decomposition Tree (PDT) data structure to be used in conjunction with scheduling and clustering techniques to analyse alternative pipelines. They also considered heterogeneous data-level parallelism where data with differing sizes is processed by multiple invocations of the same task in parallel. Unlike [80], the work in [81] evaluated different mappings of a task graph on a fixed MPSoC architecture using variants of list scheduling algorithm including As Soon As Possible (ASAP), As Late As Possible (ALAP), Earliest Deadline First (EDF) and Least Laxity (LL).

Mapping and scheduling in the context of a streaming language, StreamIt [5], has also been explored in several works. Carpenter et al. [82] proposed an iterative heuristic for partitioning a task graph and allocation of the tasks to processors in a heterogeneous MPSoC. Like [70, 74], an initial partitioning of the task graph was obtained which was then successively refined using merging of tasks, movement of bottleneck tasks, creation of new convex and connected tasks, and reallocation of tasks. Hashemi et al. [83, 84] proposed exact and approximate algorithms for mapping of task graphs onto homogeneous [83] and heterogeneous [84] MPSoCs containing only two processors. In [83], they also proposed a heuristic for MPSoCs with more than two homogeneous processors. Their algorithms are based on graph transformations to cut the task graph so that the throughput is maximised, considering interprocessor communication costs and memory sizes. Experiments with a number of StreamIt benchmarks were conducted to compare the proposed mapping techniques to StreamIt's built-in mapper [6].

Stuijk et al. [85] studied resource allocation problem under a throughput constraint in a tiled heterogeneous MPSoC when multiple multimedia applications, represented as SDFs, need to be mapped on a fixed number of tiles. The proposed heuristic sorts the tasks based on their impact on the throughput, and then greedily assigns tasks one at a time to balance the workload across all the tiles. Once a mapping is available, if it is possible, the tasks are moved around tiles to further balance the workload.

Several researchers have also looked at energy- and memory-aware mapping and scheduling of tasks on heterogeneous MPSoCs. Kim et al. [86] explored heterogeneous scheduling policies while minimising energy consumption through power management (that is, turn off a processor when it is idle to reduce energy consumption). Their heuristic approach generates a set of initial mappings of the application task graph onto the processors. Then, for each of those initial solutions, the heuristic explores scheduling policies per processor with their power management to produce Pareto-optimal design points representing energy consumption and area footprint trade-off. Yang et al. [87] proposed an approximation algorithm to partition and map a task graph onto a heterogeneous MPSoC where each processor employed Dynamic Voltage Scaling (DVS) to control its energy consumption. The objective was to minimise the energy consumption (both dynamic and leakage) of the MPSoC. This work was extended in [88] for a heterogeneous MPSoC with an arbitrary number of processors. Some of the other works on energy-efficient task graph partitioning and mapping have been reported in [89–92]. Unlike the above mentioned works, Ozturk et al. [93] proposed a compiler based approach to exploit data- and task-level parallelisms of multimedia applications in a heterogeneous MPSoC with DVFS for minimisation of energy consumption.

Bathen et al. [94, 95] explored memory-aware pipelined scheduling of multimedia applications on a homogeneous MPSoC, where the processors were equipped with scratchpad memories and were connected to a shared memory through a shared bus. They considered data allocation during the mapping and pipelined scheduling of tasks with the objective of minimising data transfers and power consumption. Salamy et al. [96, 97] explored the problem of mapping and pipelined scheduling of tasks and partitioning the available scratchpad memory among the processors simultaneously with the objective of maximising throughput. They compared their heuristic approach with an ILP formulation proposed in [55], and the results indicated that the heuristic solutions were off by a maximum of 13 % from optimal solutions.

In contrast to deterministic heuristic approaches described above, researchers have also applied random heuristic approaches comprising of Tabu Search (TS) [98], Simulated Annealing (SA) [99], Genetic Algorithm (GA) [100] and Evolutionary Algorithm (EA; parent class of GA) [101] to design space exploration and optimisation of MPSoCs. Ercan et al. [102] compared TS, SA and GA with list scheduling heuristics through the minimisation of throughput and deduced that solutions found by these random heuristics were better than those found by deterministic heuristics. However, in [102], the mapping and scheduling of tasks were done on a homogeneous MPSoC. Tumeo et al. [103, 104] and Branca et al. [105] proposed TS, SA, GA, Ant Colony Optimisation (ACO) and Bayesian Optimisation Algorithm (BOA) based heuristics to map multimedia applications with pipelined execution on the heterogeneous MPSoC proposed in [36]. The BOA based heuristic outperformed others, but with a higher running time. Yang et al. [106] studied the classical problem of mapping and scheduling a task graph on a heterogeneous MPSoC with the consideration of data-, task- and pipeline-level parallelisms. They used a Quantum-inspired Evolutionary Algorithm (QEA) to either maximise throughput or minimise MPSoC area footprint (processors and pipeline buffers) under task deadlines. The solutions from QEA were compared to the optimal solutions from ILP.

Multiple objective functions and constraints are often imposed during design space exploration of heterogeneous MPSoCs, resulting in a need for multi-objective optimisation of these MPSoCs. EAs have emerged as an attractive heuristic approach to multi-objective optimisation because they are able to quickly find near-Pareto-optimal fronts of large design spaces. Application of EAs to multi-objective optimisation of MPSoCs include classical task graph mapping and scheduling problems [107, 108], exploration of application mapping [109], and simultaneous optimisation of data allocation to memories and bus architecture [110].

Heterogeneous MPSoCs built from ASIPs exploit customisation to balance the execution time across all the processors. For example, processors executing computationally intensive tasks (over-utilised processors) can be augmented with custom instructions, hardware accelerators, special register files and the like, while functional units can be removed and cache sizes reduced in processors with less intensive tasks (under-utilised processors) so as to balance the execution time across all the processors while reducing area footprint. Customisation of ASIPs adds another dimension to the design space; therefore, several researchers have studied design space exploration of ASIP based MPSoCs in particular.

Givargis et al. [111] proposed a generic system-level design space exploration heuristic for an MPSoC, consisting of parameterisable components such as ASIPs to find Pareto-optimal fronts. Their idea was to build a parameter interdependency graph to capture the interactions between all the parameters in the MPSoC. For example, cache size and cache line size depend on each other but the voltage levels of a processor do not depend on its cache configuration. Their heuristic first finds Pareto-optimal design points in graph clusters (local Pareto-optimal front), which are then used to find Pareto-optimal design points of the MPSoC (global Pareto-optimal front). Experiments with several multimedia applications showed that the proposed two phase, iterative exploration heuristic significantly reduced the exploration time.

In contrast to [111], Sun et al. [112] studied customisation of ASIPs in heterogeneous MPSoCs in particular. They motivated the need for selection of custom instructions simultaneously with mapping and pipelined scheduling of an application's tasks on a multi-ASIP MPSoC. Their aim was to minimise the execution time of the application on an MPSoC with a fixed number of processors while the area footprint of custom instructions did not exceed the given constraint. The authors proposed an iterative heuristic approach that initially assigns and schedules tasks on the processors. Then, the processors on the critical path are augmented with custom instructions to reduce execution time while the area footprint of the processors on non-critical paths is reduced by relaxation of the already added custom instructions. Experiments with several multimedia applications showed that customised MPSoCs built with ASIPs outperformed their homogeneous counterparts in performance by up to 2.9 ×.

Exploitation of task- and pipeline-level parallelisms of multimedia applications by ASIP based pipelined MPSoCs has also been explored. Karkowski et al. [113] exploited pipeline-level parallelism by mapping one of the main loops of an application onto a pipeline of ASIPs. Multimedia applications, written in C language, were parsed to derive a graph where vertices represented statements and edges represented data dependencies. A sub-graph representing a particular main loop of the application was then selected to be compacted by merging some of the vertices (that is, vertices representing combinable statements). The reduced sub-graph was mapped on a pipeline with fixed number of ASIPs using modulo scheduling and first-fit decreasing algorithms. Experiments with a frequency tracking system showed that the pipeline of ASIPs can provide high throughput, but the computational workload need to be evenly distributed among the pipeline stages. The authors attempted workload balancing through balanced partitioning of the reduced sub-graph. Karkowski et al. extended this work in [114] where data- and pipeline-level parallelisms were considered together, in addition to arbitrary number of ASIPs in the pipelined MPSoC. A design space exploration algorithm was proposed to generate Pareto-optimal front representing performance-area tradeoff with varying number of

ASIPs. However, their work did not consider the selection of custom instructions for the ASIPs.

Shee et al. [48] explored the design space of a pipelined MPSoC where each processor had a number of configurations with performance-area tradeoffs. These processor configurations included differing custom instructions and cache configurations. The authors proposed a heuristic to select one configuration per processor with the objective of maximising pipelined MPSoC's execution time improvement per area increase ratio compared to a uniprocessor system. Their heuristic employs an analytical model to estimate the execution time of the pipelined MPSoC for a given set of processor configurations and a pruning technique. The heuristic first selects the configuration with maximum performance per area ratio for the critical processor, and then relaxes the configurations of other processors based on the critical processor's configuration to reduce the MPSoC's area footprint. Shee et al. did not consider performance constraints that are typical of real-time multimedia applications.

The work in [48] assumed a particular mapping of multimedia application's tasks on the ASIPs and then generated processor configurations (custom instructions and cache configurations) according to the mapped tasks. On the other hand, Chen et al. [115] explored simultaneous mapping and custom instruction selection with variable number of ASIPs in the pipelined MPSoC. In addition, their aim was to minimise the MPSoC's area footprint under a throughput constraint. They proposed a dynamic programming algorithm and compared it to an ILP formulation. The throughput of a multimedia application may vary because of the data-dependent nature of such applications. Therefore, Bordoli et al. [116] considered variations in processor latencies during the selection of custom instructions. They proposed a heuristic based on the branch and bound technique to select custom instructions for processors with the objective of minimum variation in throughput under an area footprint constraint for the custom instructions. Like [48], they assumed that the application was mapped to the pipelined MPSoC beforehand.

2.3.3 (Semi-) Automated Frameworks

Design space exploration approaches described above are often deployed as part of (semi-) automated frameworks to ease whole or part of an MPSoC's design process. This section provides a review of some of the design flows and design automation frameworks.

In [117], a programmer-driven, semi-automatic framework was proposed to generate different parallel specifications of an application from its sequential C code. The authors proposed six different code transformations: loop splitting; cumulative access type analysis; partitioning vector dependants; breaking composite structures; synchronising dependant variables; and, variable re-scoping. These code transformations are implemented using automated phases, but the decision to apply them is left to the programmer. The MPA framework [118] also generates parallel code from a sequential C code. A programmer marks the sections of the C code that need to be parallelised, and the analysis of parallel sections and code generation phases are automated.

Tournavitis et al. [119] proposed a profile-driven parallelisation framework where dynamic data flow analysis rather than mere static analysis was used to extract parallel loops from sequential C code. They also relied on the programmer to decide the loops that should be parallelised at the end. Ceng et al. [120] proposed MAPS framework to extract coarse-grained parallelism from sequential C code by transforming it to a weighted statement control flow graph, which is annotated with profiling information to enable both static and dynamic analyses. This graph is then processed by a heuristic to cluster so-called coupled blocks to automatically generate parallel code where the granularity of the parallelisation is controlled by the programmer. This work was later improved in [121] to better guide a programmer on the granularity of the parallel tasks. Cordes et al. [122, 123] proposed an integer linear programming based framework to extract task- and pipeline-level parallelisms by representing a sequential C code in a hierarchal task graph. The granularity of the parallelisation is automatically controlled by the costs of task creation and communication between the tasks, which are provided by the programmer.

The frameworks described above focussed on parallelisation of applications for MPSoCs. There have been other frameworks that focused on system-level design automation to rapidly explore computation and communication elements in a heterogeneous MPSoC. A Simulink and SystemC based framework for hardware-software co-design of heterogeneous MPSoCs was proposed in [124, 125]. The proposed framework allows a designer to input application and MPSoC architecture at a high level of abstraction in Simulink, which is then refined to an implementation in SystemC. A Simulink Combined Algorithm/Architecture Model (CAAM) is used to capture the abstract algorithmic flow of application and the abstract hardware components of the MPSoC. From CAAM, a hardware generator produces MPSoC architecture at three abstraction levels: virtual architecture; transaction-accurate architecture; and, virtual prototype, which trade-off simulation speed with accuracy. A multithreaded code generator then generates codes of the abstract application tasks, with some memory optimisations, for the processing elements in different abstractions of the MPSoC. Differing MPSoC architectures and an application's parallel specifications can be inputted manually, but evaluated quickly because of the automated refinement steps from application specification to MPSoC implementation.

Lyonnard et al. [126] proposed a framework where generic heterogeneous MPSoC templates with parametrisable components were used to describe an MPSoC architecture. They focused on communication coprocessors to connect heterogeneous processors and automated the generation of such coprocessors from processor and communication protocol libraries. Their framework allows a designer to explore different implementations of communication coprocessors. Wolf et al. [127] proposed an interface-centric framework for design and programming of MPSoCs. A Task Transaction Level (TTL) interface was proposed to describe both applications and MPSoCs at a high abstraction level. The TTL supports different types of communication primitives with differing implementations to trade-off performance with programming simplicity. The authors proposed several source code transformations for effective use of TTL, and automated the transformation phases for quick implementation of an application on an MPSoC architecture.

The works in [128–130] proposed frameworks for the automated exploration of on-chip communication architecture and memory in MPSoCs. Lahiri et al. [128] proposed several algorithms to generate an optimised on-chip communication architecture in the presence of differing network topologies and communication protocols. Their framework decides the mapping of communication components of the MPSoC to channels in the communication architecture template, and the protocols to be used for each of those channels. The framework in [129] automatically builds a virtual architecture of an MPSoC with differing communication architectures provided in a library by the designer for quick simulation and performance evaluation. All the communication architectures are then explored exhaustively to select the one with the best performance. Pasricha et al. [130] proposed a framework for simultaneous exploration and optimisation of the communication architecture and memory in an MPSoC. Their framework would output a bus-matrix type of communication architecture with the minimum number of busses under performance and memory area constraints. In particular, the framework determines bus topology, arbitration schemes, bus speeds and buffer sizes simultaneously with mapping of the data to memories and number, size, ports and type of each memory.

Some frameworks for automated exploration of ASIP based MPSoCs have also been proposed. Wieferink et al. [131] proposed a framework for simultaneous exploration of ASIP and communication architectures in an MPSoC. Their framework uses LISA Architectural Description Language (ADL) [132] for the description of ASIP architectures and SystemC Transaction Level Modelling (TLM) to capture communication architectures. The automatically generated ASIP simulators are interfaced with TLM communication models in the framework to allow exploration at different abstraction levels, with automated refinement from one abstraction level to the other. An ad hoc, iterative exploration approach, driven by the designer, is employed to successively improve the ASIP and communication architecture.

Angiolini et al. [133] proposed a framework for exploration of ASIPs, caches, memories and communication architectures in an MPSoC. Their framework integrates a commercial ASIP platform (LISATek [134]) with an academic MPSoC environment (MPARM [135]) because both these tools are based on SystemC and MPARM supports plug-and-play functionality. LISATek allows exploration of ASIPs while MPARM allows exploration of memory hierarchies such as scratchpad memories and caches, and communication architectures like shared busses, crossbars and NoCs. Their framework left design space exploration methodology to be implemented by the designer.

Another ASIP based MPSoC exploration framework was proposed in [136], built around Tensilica's Xtensa LX [20] processors. The MPSoC is described in XML to be used by automated tools to generate simulation models for the hardware components of the MPSoC, and separate executables for each processor from the C codes. The framework then allows simulation of the MPSoC at either the cycle-accurate or functional level, while trading-off simulation speed with accuracy. A designer can explore MPSoCs with different number and types of processors, buffer sizes and so on within several hours, although the MPSoC architectures need to be inputted manually.

The frameworks described above focused on automation of some of the design phases of a heterogeneous MPSoC. The following paragraphs review more complete, state-of-the-art frameworks for heterogeneous MPSoCs [137]. Metropolis [138] provides a modelling and simulation environment based on the Platform Based Design (PBD) paradigm [139]. The PBD simplifies system-level design by constraining the MPSoC to a fixed architectural template so that the design problem reduces to mapping of an application onto the MPSoC template. Metropolis uses a meta-model language to capture application functionality and the MPSoC platform. The meta-model employs an event-based execution model where processes communicate with each other through channels. For a given application and architectural template, synthesis is performed by mapping the application onto the MPSoC where different phases of the synthesis such as parsing of meta-models, generation of SystemC simulation models, scheduling and so on are automated.

Koski [140] provides a framework for the following: modelling of an application; automated MPSoC design space exploration; and, automated synthesis, programming and prototyping on FPGA of the selected MPSoC design. An application is described in a UML model for mapping onto a bus-based MPSoC architecture, which is constructed from a library of components. The UML interface transforms the application and MPSoC descriptions to an abstract level for fast exploration. A two step MPSoC architecture exploration approach is employed where the designer can specify performance, area and power constraints. Once an MPSoC design is selected, generation of code for application tasks, RTL description of components, and integration of RTOS are automated for implementation on an FPGA.

PeaCE [141] is an extension of Ptolemy II [142], and provides a hardware-software co-design framework from functional simulation to MPSoC prototyping. PeaCE uses extended synchronous data flow graphs and finite state machines to model data flow and control flow of multimedia applications. The MPSoC architecture consists of a number of processors and synthesise-able IP cores, which are connected through a communication architecture. A two step design space exploration is used. The first step explores the selection of processing elements and mapping of application tasks on those elements. The second step involves the exploration of the communication architecture such as bus and memory allocation. After design space exploration, the chosen MPSoC designs can be prototyped on an FPGA. Another enhancement to PeaCE, named HOPES [143], was proposed to ease the programming of MPSoCs. HOPES introduces the Common Intermediate Code (CIC) model to capture both the application and the MPSoC architecture. The CIC model can be either written manually or generated automatically from PeaCE models. The CIC translator in HOPES transforms the model to optimised software codes for processors and interface code for IP cores with the scheduling of application tasks.

Daedalus [144, 145] provides a highly automated framework for system-level exploration, synthesis, programming and prototyping of MPSoCs by combining KPNgen [146], Sesame [147] and ESPAM [148] tools. Daedalus uses the Kahn Process Network (KPN) as the model of computation and composable, heterogeneous

MPSoCs (created from a library of components) where the processing elements communicate through distributed memories as the implementation platform. The KPN of an application is either derived manually or automatically by utilising KPNgen if the application's sequential C code is specified as a so-called static affine nested loop program. KPNgen can also use automated source level transformations to produce different KPNs of an application. The generated KPNs are then used by Sesame to perform design space exploration of mapping KPNs, scheduling processes of KPNs, and communication and computation components in the MPSoC platforms. Sesame trades-off simulation speed with accuracy by the use of either high- or low-level architectural models. A set of promising KPNs and MPSoC platforms from Sesame can be passed to ESPAM for prototyping on an FPGA. ESPAM automatically generates C code of the processes in the KPN and synthesise-able VHDL of the MPSoC platform from RTL models in the component library. Daedalus allows quick exploration of differing application mappings and schedules, and differing MPSoC platforms because of the automated design trajectory from application specification to implementation. Recently, Daedalus^{RT} has been proposed in [149] for mapping and scheduling of multiple multimedia applications with hard real-time throughput constraints.

SCE [150] is another framework for automated implementation of an application on a heterogeneous MPSoC. Unlike previous frameworks, SCE is based on SpecC [151] system-level design language and provides an interactive, user-driven GUI. In SCE, an application is described as a hierarchical state machine while the MPSoC platform is built from a library of components. SCE employs a Specify-Explore-Refine approach to implement the specified application onto a predefined MPSoC platform using a predefined mapping. The "explore" step consists of four types of explorations: architectural (selection of processing elements and memories, mapping of application tasks and data); scheduling of application tasks on selected processing elements; communication architecture (selection of buses, communication elements and their connectivity); and, mapping of channels onto the communication architecture. SCE uses a plug-in approach for inclusion of user-defined exploration and optimisation algorithms. In the last step, SCE automatically refines the selected design point from the "explore" step by generating RTL models of the hardware components and executables of the application's tasks.

SystemCoDesigner [152] provides a framework to automatically map and schedule multimedia applications onto a heterogeneous MPSoC where applications consist of actors, which communicate through channels. After the specification of the application and MPSoC template in SystemC by the designer, the SystemCoDesigner generates hardware accelerators for the actors and adds those accelerators to a component library made up of processors, IP cores, buses and memories. For quick evaluation of differing implementations of the MPSoC template, SystemCoDesigner translates the MPSoC into a so-called virtual architecture. Unlike previous frameworks, for design space exploration, SystemCoDesigner transforms the input SystemC model into a pseudo-Boolean formula and uses multi-objective evolutionary algorithms. The Pareto-optimal designs from the exploration phase can be automatically prototyped on an FPGA.

2.4 Run-Time Adaptability

The optimisation and exploration frameworks described above are typically used at design-time to optimise heterogeneous MPSoCs under worst-case parameters so that these MPSoCs, when deployed, can deliver the performance required of them at all times. Design-time optimised MPSoCs lack adaptability, and thus result in inefficient resource utilisation and increased energy consumption under a dynamic environment. In a dynamic environment, run-time adaptability is an attractive option to improve the resource utilisation and energy efficiency of heterogenous MPSoCs. There is a plethora of work on run-time adaptability in MPSoCs; however, this section provides an overview of only some of the run-time management techniques.[1]

Most of the run-time management techniques adapt the MPSoC under dynamic workload by exploiting task migration, mapping and scheduling to increase utilisation of the processors. In addition, these techniques use Dynamic Voltage and Frequency Scaling (DVFS) or multiple power states to reduce energy consumption by reducing frequency-voltage levels of under-utilised processors and/or by transitioning idle processors to sleep states.

Schranzhofer et al. [154] addressed the problem of selection of processing elements in a heterogeneous MPSoC, and the mapping of multiple applications onto the selected processing elements under different application scenarios (representing dynamic workload). Their objective was to minimise the average power consumption of the heterogeneous MPSoC by improving the utilisation of processing elements. They proposed a two-step solution: firstly, an approach to select processing elements and to compute a static schedule of the tasks was introduced; and secondly, a set of promising static schedules for all the application scenarios was computed offline and stored in the MPSoC to be used by the run-time manager during differing application scenarios.

Since the overhead of run-time management techniques has to be low, like [154], Couvreur et al. [155] also proposed a two step approach. A multi-objective design space exploration is performed to obtain the Pareto-optimal design points, which are stored in the MPSoC for use at run-time. Then, one of the Pareto-optimal design points is selected by the run-time manager, considering run-time resource utilisation of the MPSoC. The objective of their design space exploration was to minimise energy consumption of the MPSoC under a performance constraint, where memory usage, processor frequencies, communication bandwidth, and the like constituted MPSoC's resources. In addition, the authors used differing parallelisations of the application based upon workload variations as the dynamic factor. Other similar works where design-time decisions are coupled with run-time management techniques are reported in [156].

[1] Dynamically reconfigurable systems use partial reconfiguration characteristic of FPGAs to adapt their hardware to application demands at run-time. Since this monograph targets heterogenous MPSoCs with non-reconfigurable hardware, literature on dynamic reconfigurable systems is not covered here. However, interested readers are directed to [153], where the authors report some of the recent adaptive, reconfigurable systems and their run-time management techniques.

In situations where the dynamic nature of the system cannot be modelled at design-time (such as the variations in workload due to input data), more advanced run-time management techniques are deployed. These techniques are based upon design-time analytical analyses, run-time monitoring and run-time prediction of the MPSoC's resource utilisation, power consumption, and the like. In [157], a run-time heuristic to select frequency-voltage levels for components in a Globally Asynchronous Locally Synchronous (GALS) system with frequency-voltage islands was proposed. The proposed heuristic predicts the execution cycles for the next epoch based upon the predicted and run-time monitored, actual execution cycles of the previous epochs. Then, the predicted execution cycles are used to select an appropriate frequency-voltage level for a component.

Isci et al. [158] proposed a global power manager to apply DVFS in an MPSoC, where its workload changes at run-time. Like [157], they proposed a heuristic which monitors performance and power of the MPSoC during an epoch, and then uses the measured values to predict the performance and power for the next epoch. The heuristic uses the predictions to select frequency-voltage level that will maximise performance under a power budget. Puschini et al. [159] proposed a run-time management technique, inspired from game theory, to decide the frequency-voltage levels of processors in an MPSoC. The decisions are made locally for each of the processors with the objective of latency minimisation under run-time varying energy constraint or energy minimisation under run-time varying latency constraint.

Unlike the above mentioned works, Molnos et al. [160] proposed an OS-level run-time technique to select frequency-voltage levels of processors in an MPSoC. Their heuristic conservatively exploits the slack that occurs at run-time (due to the dynamic workload) by allocating that slack to a ready task, and then by lowering the ready task's frequency-voltage level. The heuristic determines the slack at run-time by monitoring the execution of the tasks, and minimises the energy consumption under a performance constraint.

Huang et al. [161] proposed a run-time task mapping and scheduling technique to maximise resource utilisation of an MPSoC under a performance constraint. A four step heuristic was proposed: firstly, the application deadline (performance constraint) is translated to individual task deadlines with the objective of maximal scheduling; secondly, adaptive task mapping and clustering is performed with the objective of maximising resource utilisation and minimising communication costs respectively, considering the run-time feedback from the MPSoC; thirdly, local scheduling of tasks on each of the processors; and lastly, bandwidth allocation in the NoC. Their adaptive task mapping was shown to outperform traditional task mapping, which did not consider run-time feedback from the MPSoC about its resource utilisation.

In [162], the problem of task mapping and scheduling of an application graph exhibiting dynamic workload on an MPSoC with DVFS was addressed. The dynamic workload of an application is due to the conditional branches in the graph as some of the tasks may not be executed at run-time. The branches are associated with probabilities which are populated at run-time by monitoring branch selections. Based on the captured history of the branches, a run-time heuristic schedules the tasks and selects frequency-voltage levels for the processors. The heuristic is called every time a branch probability changes by more than a predetermined threshold.

Huang et al. [163] addressed the problem of selecting power states of a device (such as an MPSoC) in the presence of dynamic event streams in order to minimise its average power consumption. They used real-time calculus to model the event streams and hence to predict the future arrival of events. Based on a combination of past and predicted events, decisions are made on the power state of the device. A power state is used only when the device is predicted to be idle for a long enough period to amortise the overhead of transitioning to that particular power state.

Unlike the above mentioned works which primarily focused on maximising resource utilisation or minimising energy consumption, Coskun et al. [164] proposed a proactive, run-time thermal management technique for MPSoCs. They used the autoregressive moving average model for prediction of future temperature, based on the temperature measurements of the past. In addition, they applied sequential probability ratio test to predict when the predictions of moving average predictor will significantly drift from actual measurements so that the predictor can be adapted in advance. Experiments were conducted with task migration and the Dynamic Voltage Scaling (DVS) enabled run-time thermal management to illustrate that the proposed proactive technique significantly reduced the frequency of hotspots compared to reactive techniques. On the other hand, Ebi et al. [165] used an agent based run-time thermal management technique to proactively avoid potential hotspots that may develop in future.

Machine learning has also been used in prior research to learn at run-time from the history of an application's execution so as to enable future predictions. Ge et al. [166] applied machine learning to thermal management while Tan et al. [167] applied it to power management in MPSoCs. Several techniques have also been proposed for run-time management of communication in NoC based MPSoCs. The authors of [168] studied the problem of task mapping of an application at run-time with the objective of minimisation of NoC congestion. They proposed several heuristics based on first-free neighbour, nearest-neighbour, communication path load, etc. and compared their effectiveness in improvement in channel load, packet latency and execution time. Al Faruque et al. [169] proposed a distributed, agent based run-time task mapping technique for a similar problem.

Run-time adaptability has also been used in pipelined MPSoCs to adapt them under dynamic workload. In pipelined MPSoCs, the variations in workload are typically due to the data-dependent behaviour of multimedia applications, resulting in an unbalancing of the pipeline stages at run-time. Therefore, the authors of [45–47] introduced per processor DVFS to reduce frequency-voltage level when a processor is under-utilised and to increase frequency-voltage level when a processor exceeds the throughput constraint. Consequently, the stages are balanced at run-time under workload variations by making their latencies almost the same, and close to the pipelined MPSoC's throughput constraint.

Guo et al. [45] proposed a feedback controller based centralised run-time manager. The run-time manager uses one detector per FIFO buffer in the pipelined MPSoC to monitor its utilisation by its producer and consumer processors. The voltage level is changed when the difference in utilisation of a FIFO buffer's producer and consumer is more than a threshold, indicating that the two processors are unbalanced. The feedback controller is triggered every time a task is executed.

Like [45], in [46, 47], a feedback controller, based upon the occupancy levels of the FIFO buffers, was proposed to select frequency-voltage levels of the processors. However, the authors of [46, 47] proposed distributed, more fine-grained, both linear and non-linear controllers where the feedback control policy is executed in each of the processors of the pipelined MPSoC. The proposed controller can be triggered either after a fixed time interval or adaptively based on the occupancy level of the FIFO buffer. Every time a controller is triggered, it measures the current occupancy level of the FIFO buffer and compares it to the desired and previous occupancy levels of the FIFO buffer. If the error in occupancy level is within a threshold, then the frequency-voltage level is unchanged. On the other hand, if the error in occupancy level is negative (that is, the FIFO buffer is being filled slowly), then the frequency-voltage level is increased. Otherwise, the frequency-voltage level is decreased. Although feedback controllers provide run-time adaptability in pipelined MPSoCs, they are reactive in nature rather than proactive as they do not utilise any form of prediction. Therefore, the controller only acts when a performance penalty has occurred, instead of forecasting such a penalty and acting in advance. Proactive techniques are typically required when the workload variations are sudden, due to input data dependence, which is the case with multimedia applications.

2.5 Summary

This chapter opened with a survey of homogeneous and heterogenous MPSoCs used for multimedia. The chapter then focused on design space exploration of heterogeneous MPSoCs. Both exact and heuristic approaches typically utilised during design space exploration were discussed. An overview of several (semi-) automated frameworks that can ease and speed up the design and prototyping of heterogeneous MPSoC was also provided. Finally, the chapter focussed on run-time adaptability in heterogeneous MPSoCs under dynamic environments. In summary, the chapter provided the necessary survey of the existing design-time and run-time optimisation techniques for heterogeneous MPSoCs in general and pipelined MPSoCs in particular.

References

1. U. Kapasi, W. Dally, S. Rixner, J. Owens, B. Khailany, The imagine stream processor. in Computer Design: VLSI in Computers and Processors, 2002. Proceedings. 2002 IEEE International Conference on, pp. 282–288, 2002.
2. J. H. Ahn, W. J. Dally, B. Khailany, U. J. Kapasi, and A. Das, "Evaluating the imagine stream architecture", in Proceedings of the 31st annual international symposium on Computer architecture, ISCA '04, (Washington, DC, USA), pp. 14-, IEEE Computer Society, 2004.

3. M. Taylor, J. Kim, J. Miller, D. Wentzlaff, F. Ghodrat, B. Greenwald, H. Hoffman, P. Johnson, J.-W. Lee, W. Lee, A. Ma, A. Saraf, M. Seneski, N. Shnidman, V. Strumpen, M. Frank, S. Amarasinghe, and A. Agarwal, "The raw microprocessor: a computational fabric for software circuits and general-purpose programs", Micro, IEEE, vol. 22, pp. 25–35, mar/apr 2002.
4. M. B. Taylor, W. Lee, J. Miller, D. Wentzlaff, I. Bratt, B. Greenwald, H. Hoffmann, P. Johnson, J. Kim, J. Psota, A. Saraf, N. Shnidman, V. Strumpen, M. Frank, S. Amarasinghe, and A. Agarwal, "Evaluation of the raw microprocessor: An exposed-wire-delay architecture for ilp and streams", in ISCA '04: Proceedings of the 31st annual international symposium on Computer architecture, (Washington, DC, USA), p. 2, IEEE Computer Society, 2004.
5. W. Thies, M. Karczmarek, and S. P. Amarasinghe, "Streamit: A language for streaming applications", in Proceedings of the 11th International Conference on Compiler Construction, CC '02, pp. 179–196, Springer-Verlag, 2002.
6. M.I. Gordon, W. Thies, S. Amarasinghe, Exploiting coarse-grained task, data, and pipeline parallelism in stream programs. SIGPLAN Not. **41**, 151–162 (Oct. 2006)
7. Tilera, "Tilepro64 multicore processor product brief". Available at http://www.tilera.com
8. A. Agarwal, L. Bao, J. Brown, B. Edwards, M. Mattina, C.-C. Miao, C. Ramey, and D. Wentzlaff, "Tile processor: Embedded multicore for networking and multimedia", in Hot Chips, 2007.
9. Intel, "Single-chip cloud computer". Available at http://www.intel.com
10. J. Howard, S. Dighe, S. Vangal, G. Ruhl, N. Borkar, S. Jain, V. Erraguntla, M. Konow, M. Riepen, M. Gries, G. Droege, T. Lund-Larsen, S. Steibl, S. Borkar, V. De, and R. Van Der Wijngaart, "A 48-core ia-32 processor in 45 nm cmos using on-die message-passing and dvfs for performance and power scaling", Solid-State Circuits, IEEE Journal of, vol. 46, pp. 173–183, jan. 2011.
11. P. Gschwandtner, T. Fahringer, and R. Prodan, "Performance analysis and benchmarking of the intel scc", in Cluster Computing (CLUSTER), 2011 IEEE International Conference on, pp. 139–149, sept. 2011.
12. T. G. Mattson, M. Riepen, T. Lehnig, P. Brett, W. Haas, P. Kennedy, J. Howard, S. Vangal, N. Borkar, G. Ruhl, and S. Dighe, "The 48-core scc processor: the programmer's view", in Proceedings of the 2010 ACM/IEEE International Conference for High Performance Computing, Networking, Storage and Analysis, SC '10, (Washington, DC, USA), pp. 1–11, IEEE Computer Society, 2010.
13. C. Clauss, S. Lankes, P. Reble, and T. Bemmerl, "Evaluation and improvements of programming models for the intel scc many-core processor", in High Performance Computing and Simulation (HPCS), 2011 International Conference on, pp. 525–532, july 2011.
14. R. Kumar, D. M. Tullsen, P. Ranganathan, N. P. Jouppi, and K. I. Farkas, "Single-isa heterogeneous multi-core architectures for multithreaded workload performance", SIGARCH Comput. Archit. News, vol. 32, pp. 64-, Mar. 2004.
15. W. Wolf, A. Jerraya, and G. Martin, "Multiprocessor system-on-chip (mpsoc) technology", Computer-Aided Design of Integrated Circuits and Systems, IEEE Transactions on, vol. 27, pp. 1701–1713, oct. 2008.
16. Freescale, "C-5 network processor". Available at http://www.freescale.com
17. Intel, "Ixp network processors". Available at http://www.intel.com
18. W. Eatherton, "Silicon packet processor", in Symposium on Architecures for Networking and Communications Systems, 2005.
19. Cisco, "The cisco quantumflow processor: Cisco's next generation network processor". Available at http://www.cisco.com
20. Tensilica, "Xtensa Customizable Processor". http://www.tensilica.com
21. S. Dutta, R. Jensen, and A. Rieckmann, "Viper: A multiprocessor soc for advanced set-top box and digital tv systems", Design Test of Computers, IEEE, vol. 18, pp. 21–31, sep-oct 2001.
22. NXP, "Nxp trimedia architecture". Available at http://www.nxp.com, 2012
23. Nxp Semiconductors, "Nexperia media processor". Available at: http://www.nxp.com
24. STMicroelectronics, "Nomadik application processor". Available at http://www.st.com

25. STMicroelectronics and Ericsson, “Novathor platform for smartphones and tablets”. Available at http://www.stericsson.com
26. ARM, “Arm processors”. Available at http://www.arm.com
27. ARM, “Mail multimedia hardware”. Available at http://www.arm.com
28. ARM, “Neon general-purpose simd enginer”. Available at http://www.arm.com
29. H. P. Hofstee, “Power efficient processor architecture and the cell processor”, in Proceedings of the 11th International Symposium on High-Performance Computer Architecture, pp. 258–262, IEEE Computer Society, 2005.
30. NVIDIA, “Tegra multiprocessor architecture”. Available at http://www.nvidia.com/
31. Texas Instruments, “Omap mobile processors”. Available at http://www.ti.com/
32. Imagination Technologies, “Powervr graphics”. Available at http://www.imgtec.com
33. M. Strik, A. Timmer, J. van Meerbergen, G.-J. van Rootselaar, Heterogeneous multiprocessor for the management of real-time video and graphics streams. Solid-State Circuits, IEEE Journal of **35**(11), 1722–1731 (Nov 2000)
34. A. Beric, R. Sethuraman, C. Pinto, H. Peters, G. Veldman, P. van de Haar, and M. Duranton, “Heterogeneous multiprocessor for high definition video”, Consumer Electronics, 2006. ICCE ’06. 2006 Digest of Technical Papers. International Conference on, pp. 401–402, 7–11 Jan. 2006.
35. D. Wu, P. Karlstorm, J. Eilert, A. Ehlair, and D. Liu, “Mediadsp: An application specific heterogeneous multiprocessor soc”, in Proc. Swedish System-on-Chip Conf. (SSoCC), 2006.
36. A. Tumeo, M. Branca, L. Camerini, M. Ceriani, M. Monchiero, G. Palermo, F. Ferrandi, and D. Sciuto, “Prototyping pipelined applications on a heterogeneous fpga multiprocessor virtual platform”, in ASP-DAC ’09: Proceedings of the 2009 Asia and South Pacific Design Automation Conference, 2009.
37. C. May, E. Silha, R. Simpson, H. Warren, and C. International Business Machines, Inc., eds., The PowerPC architecture: a specification for a new family of RISC processors. San Francisco, CA, USA: Morgan Kaufmann Publishers Inc., 1994.
38. Xilinx, “Microblaze soft core”. Available at http://www.xilinx.com
39. E. Hansson, J. Sohl, C. Kessler, and D. Liu, “Case study of efficient parallel memory access programming for the embedded heterogeneous multicore dsp architecture epuma”, in Complex, Intelligent and Software Intensive Systems (CISIS), 2011 International Conference on, pp. 624–629, 30 2011-july 2 2011.
40. H. Park, Y. Park, and S. Mahlke, “Polymorphic pipeline array: a flexible multicore accelerator with virtualized execution for mobile multimedia applications”, in Proceedings of the 42nd Annual IEEE/ACM International Symposium on Microarchitecture, MICRO 42, (New York, NY, USA), pp. 370–380, ACM, 2009.
41. Tensilica, “Diamond processor”. Available at http://www.tensilica.com
42. S. L. Shee, A. Erdos, and S. Parameswaran, “Heterogeneous multiprocessor implementations for jpeg: a case study”, in CODES+ISSS ’06: Proceedings of the 4th international conference on Hardware/software codesign and system synthesis, (New York, NY, USA), pp. 217–222, ACM, 2006.
43. S.L. Shee, A. Erdos, S. Parameswaran, Architectural exploration of heterogeneous multiprocessor systems for jpeg. International Journal of Parallel Programming **36**(1), 140–162 (2008)
44. R. Hameed, W. Qadeer, M. Wachs, O. Azizi, A. Solomatnikov, B.C. Lee, S. Richardson, C. Kozyrakis, M. Horowitz, Understanding sources of ineffciency in general-purpose chips. Commun. ACM **54**, 85–93 (Oct. 2011)
45. H. Guo and S. Parameswaran, “Balancing system level pipelines with stage voltage scaling”, in Proceedings of the IEEE Computer Society Annual Symposium on VLSI: New Frontiers in VLSI Design, ISVLSI ’05, 2005.
46. S. Carta, A. Alimonda, A. Pisano, A. Acquaviva, L. Benini, “A control theoretic approach to energy-efficient pipelined computation in mpsocs”, ACM Trans. Embedded, Comput. Syst., 6(4), 2007.

47. A. Alimonda, S. Carta, A. Acquaviva, A. Pisano, and L. Benini, "A feedback-based approach to dvfs in data-flow applications", IEEE Trans. on CAD of Integrated Circuits and Systems, vol. 28, no. 11, pp. 1691–1704, 2009.
48. S. L. Shee and S. Parameswaran, "Design methodology for pipelined heterogeneous multiprocessor system", in DAC '07: Proceedings of the 44th annual conference on Design automation, pp. 811–816, 2007.
49. H. Javaid, A. Ignjatovic, S. Parameswaran, Rapid design space exploration of application specific heterogeneous pipelined multiprocessor systems. Trans. Comp.-Aided Des. Integ. Cir. Sys. **29**, 1777–1789 (November 2010)
50. T. H. Cormen, C. E. Leiserson, R. L. Rivest, and C. Stien, Introduction to Algorithms. CRC Press, illustrated ed., 2002.
51. J. DeSouza-Batista and A. Parker, "Optimal synthesis of application specific heterogeneous pipelined multiprocessors", Application Specific Array Processors, 1994. Proceedings., International Conference on, pp. 99–110, 22–24 Aug 1994.
52. M. Schwiegershausen and P. Pirsch, "A formal approach for the optimization of heterogeneous multiprocessors for complex image processing schemes", in EURO-DAC '95/EURO-VHDL '95: Proceedings of the conference on European design automation, (Los Alamitos, CA, USA), pp. 8–13, IEEE Computer Society Press, 1995.
53. B. K. Dwivedi, A. Kumar, and M. Balakrishnan, "Synthesis of application specific multiprocessor architectures for process networks", in VLSID '04: Proceedings of the 17th International Conference on VLSI Design, (Washington, DC, USA), p. 780, IEEE Computer Society, 2004.
54. S.-R. Kuang, C.-Y. Chen, and R.-Z. Liao, "Partitioning and pipelined scheduling of embedded system using integer linear programming", in ICPADS '05: Proceedings of the 11th International Conference on Parallel and Distributed Systems - Workshops (ICPADS'05), (Washington, DC, USA), pp. 37–41, IEEE Computer Society, 2005.
55. V. Suhendra, C. Raghavan, and T. Mitra, "Integrated scratchpad memory optimization and task scheduling for mpsoc architectures", in Proceedings of the 2006 international conference on Compilers, architecture and synthesis for embedded systems, CASES '06, (New York, NY, USA), pp. 401–410, ACM, 2006.
56. R. Banakar, S. Steinke, B.-S. Lee, M. Balakrishnan, and P. Marwedel, "Scratchpad memory: design alternative for cache on-chip memory in embedded systems", in Proceedings of the tenth international symposium on Hardware/software codesign, CODES '02, (New York, NY, USA), pp. 73–78, ACM, 2002.
57. C. Ostler and K. Chatha, "An ilp formulation for system-level application mapping on network processor architectures", in Design, Automation Test in Europe Conference Exhibition, 2007. DATE '07, pp. 1–6, april 2007.
58. Y. Yi, W. Han, X. Zhao, A. Erdogan, and T. Arslan, "An ilp formulation for task mapping and scheduling on multi-core architectures", in Design, Automation Test in Europe Conference Exhibition, 2009. DATE '09., pp. 33–38, april 2009.
59. H. Yang and S. Ha, "Ilp based data parallel multi-task mapping/scheduling technique for mpsoc", in SoC Design Conference, 2008. ISOCC '08. International, vol. 01, pp. I-134 -I-137, nov. 2008.
60. T. Suleyman, M. Nazanin, T. K. Mahmut, and O. Ozcan, "An ilp formulation for task scheduling on heterogeneous chip multiprocessors", in ISCIS, pp. 267–276, 2006.
61. Y. Yetim, S. Malik, and M. Martonosi, "Eprof: An energy/performance/reliability optimization framework for streaming applications", in Design Automation Conference (ASP-DAC), 2012 17th Asia and South Pacific, pp. 769–774, 2012.
62. M. Ruggiero, A. Guerri, D. Bertozzi, M. Milano, L. Benini, A fast and accurate technique for mapping parallel applications on stream-oriented mpsoc platforms with communication awareness. Int. J. Parallel Program. **36**, 3–36 (Feb. 2008)
63. J. Wu, J. Williams, and N. Bergmann, "An ilp formulation for architectural synthesis and application mapping on fpga-based hybrid multi-processor soc", in Field Programmable Logic and Applications, 2008. FPL 2008. International Conference on, pp. 451–454, sept. 2008.

64. C.-L. Sotiropoulou and S. Nikolaidis, "Ilp formulation for hybrid fpga mpsocs optimizing performance, area and memory usage", in Electronics, Circuits and Systems (ICECS), 2011 18th IEEE International Conference on, pp. 748–751, dec. 2011.
65. I. Kadayif, M. Kandemir, and U. Sezer, "An integer linear programming based approach for parallelizing applications in on-chip multiprocessors", in Design Automation Conference, 2002. Proceedings. 39th, pp. 703–708, 2002.
66. Y. Choi, Y. Lin, N. Chong, S. Mahlke, and T. Mudge, "Stream compilation for real-time embedded multicore systems", in Code Generation and Optimization, 2009. CGO 2009. International Symposium on, pp. 210–220, march 2009.
67. M. Hall, J. Anderson, S. Amarasinghe, B. Murphy, S.-W. Liao, and E. Bu, "Maximizing multiprocessor performance with the suif compiler", Computer, vol. 29, pp. 84–89, dec 1996.
68. Y. Jin, N. Satish, K. Ravindran, and K. Keutzer, "An automated exploration framework for fpga-based soft multiprocessor systems", in CODES+ISSS '05: Proceedings of the 3rd IEEE/ACM/IFIP international conference on Hardware/software codesign and system synthesis, (New York, NY, USA), pp. 273–278, ACM, 2005.
69. J. Cong, G. Han, and W. Jiang, "Synthesis of an application-specific soft multiprocessor system", in FPGA '07: Proceedings of the 2007 ACM/SIGDA 15th international symposium on Field programmable gate arrays, (New York, NY, USA), pp. 99–107, ACM, 2007.
70. S. Banerjee, T. Hamada, P. Chau, R. Fellman, Macro pipelining based scheduling on high performance heterogeneous multiprocessor systems. Signal Processing, IEEE Transactions on **43**(6), 1468–1484 (1995)
71. S. Bakshi, D.D. Gajski, Component selection for high-performance pipelines. IEEE Trans. VLSI Syst. **4**(2), 181–194 (1996)
72. S. Bakshi, D.D. Gajski, Partitioning and pipelining for performance-constrained hardware/software systems. IEEE Trans. VLSI Syst. **7**(4), 419–432 (1999)
73. J. Jeon and K. Choi, "Loop pipelining in hardware-software partitioning", in Asia and South Pacific Design Automation Conference, pp. 361–366, 1998.
74. S. Ranaweera and D. Agrawal, "Scheduling of periodic time critical applications for pipelined execution on heterogeneous systems", in Parallel Processing, International Conference on, 2001, pp. 131–138, sept. 2001.
75. A. Benoit, Y. Robert, Mapping pipeline skeletons onto heterogeneous platforms. Journal of Parallel and Distributed Computing **68**(6), 790–808 (2008)
76. Y. R. Anne Benoit and E. Thierry, "On the complexity of mapping linear chain applications onto heterogeneous platforms", in, Parallel Processing Letters, 2009.
77. A. Benoit, V. Rehn-Sonigo, and Y. Robert, "Multi-criteria scheduling of pipeline workflows", in Cluster Computing, 2007 IEEE International Conference on, pp. 515–524, sept. 2007.
78. A. Benoit, H. Kosch, V. Rehn-Sonigo, and Y. Robert, "Bi-criteria pipeline mappings for parallel image processing", in ICCS '08: Proceedings of the 8th international conference on Computational Science, Part I, (Berlin, Heidelberg), pp. 215–225, Springer-Verlag, 2008.
79. V. R.-S. Anne Benoit, Harald Kosch and Y. Robert, "Multi-criteria scheduling of pipeline workflows (and application to the jpeg encoder)", in Internation Journal of High Performance Computing Applications, 2009.
80. D.-I. Ko and S. S. Bhattacharyya, "The pipeline decomposition tree: an analysis tool for multiprocessor implementation of image processing applications", in CODES+ISSS '06: Proceedings of the 4th international conference on Hardware/software codesign and system synthesis, (New York, NY, USA), pp. 52–57, ACM, 2006.
81. B. Ristau, T. Limberg, and G. Fettweis, "A mapping framework for guided design space exploration of heterogeneous mp-socs", in DATE '08: Proceedings of the conference on Design, automation and test in Europe, (New York, NY, USA), pp. 780–783, ACM, 2008.
82. P. M. Carpenter, A. Ramirez, and E. Ayguade, "Mapping stream programs onto heterogeneous multiprocessor systems", in Proceedings of the 2009 international conference on Compilers, architecture, and synthesis for embedded systems, CASES '09, (New York, NY, USA), pp. 57–66, ACM, 2009.

83. M. Hashemi and S. Ghiasi, "Throughput-driven synthesis of embedded software for pipelined execution on multicore architectures", ACM Trans. Embed. Comput. Syst., vol. 8, pp. 11:1–11:35, February 2009.
84. M. Hashemi and S. Ghiasi, "Versatile task assignment for heterogeneous soft dual-processor platforms", Computer-Aided Design of Integrated Circuits and Systems, IEEE Transactions on, vol. 29, pp. 414–425, march 2010.
85. S. Stuijk, T. Basten, M. Geilen, and H. Corporaal, "Multiprocessor resource allocation for throughput-constrained synchronous dataflow graphs", in Design Automation Conference, 2007. DAC '07. 44th ACM/IEEE, pp. 777–782, june 2007.
86. M. Kim, S. Banerjee, N. Dutt, and N. Venkatasubramanian, "Energy-aware cosynthesis of real-time multimedia applications on mpsocs using heterogeneous scheduling policies", ACM Trans. Embed. Comput. Syst., vol. 7, pp. 9:1–9:19, January 2008.
87. C.-Y. Yang, J.-J. Chen, T.-W. Kuo, and L. Thiele, "An approximation scheme for energy-efficient scheduling of real-time tasks in heterogeneous multiprocessor systems", in Proceedings of the Conference on Design, Automation and Test in Europe, DATE '09, (3001 Leuven, Belgium, Belgium), pp. 694–699, European Design and Automation Association, 2009.
88. J.-J. Chen, A. Schranzhofer, and L. Thiele, "Energy minimization for periodic real-time tasks on heterogeneous processing units", in Parallel Distributed Processing, 2009. IPDPS 2009. IEEE International Symposium on, pp. 1–12, may 2009.
89. A. Rae and S. Parameswaran, "Voltage reduction of application-specific heterogeneous multiprocessor systems for power minimisation", in Proceedings of the 2000 Asia and South Pacific Design Automation Conference, ASP-DAC '00, (New York, NY, USA), pp. 147–152, ACM, 2000.
90. R. Xu, R. Melhem, D. Mosse, "Energy-aware scheduling for streaming applications on chip multiprocessors", in Real-Time Systems Symposium, 2007. RTSS 2007. 28th IEEE, International, pp. 25–38, dec. 2007.
91. L. K. Goh, B. Veeravalli, and S. Viswanathan, "Design of fast and efficient energy-aware gradient-based scheduling algorithms heterogeneous embedded multiprocessor systems", Parallel and Distributed Systems, IEEE Transactions on, vol. 20, pp. 1–12, jan. 2009.
92. B. Virlet, X. Zhou, J. P. Giacalone, B. Kuhn, M. J. Garzaran, and D. Padua, "Scheduling of stream-based real-time applications for heterogeneous systems", in Proceedings of the 2011 SIGPLAN/SIGBED conference on Languages, compilers and tools for embedded systems, LCTES '11, (New York, NY, USA), pp. 1–10, ACM, 2011.
93. O. Ozturk, M. Kandemir, and G. Chen, "Compiler-directed energy reduction for voltage islands", Computers, IEEE Transactions on, vol. PP, no. 99, p. 1, 2011.
94. L. Bathen, N. Dutt, and S. Pasricha, "A framework for memory-aware multimedia application mapping on chip-multiprocessors", in Embedded Systems for Real-Time Multimedia, 2008. ESTImedia 2008. IEEE/ACM/IFIP Workshop on, pp. 89–94, oct. 2008.
95. L. Bathen, Y. Ahn, N. Dutt, and S. Pasricha, "Inter-kernel data reuse and pipelining on chip-multiprocessors for multimedia applications", in Embedded Systems for Real-Time Multimedia, 2009. ESTIMedia 2009. IEEE/ACM/IFIP 7th Workshop on, pp. 45–54, oct. 2009.
96. H. Salamy, J. Ramanujam, A framework for task scheduling and memory partitioning for multi-processor system-on-chip. High Performance Embedded Architectures and Compilers **5409**, 263–277 (2009)
97. H. Salamy and J. Ramanujam, "An effective solution to task scheduling and memory partitioning for multiprocessor system-on-chip", Computer-Aided Design of Integrated Circuits and Systems, IEEE Transactions on, vol. 31, pp. 717–725, may 2012.
98. F. Glover, E. Taillard, D. de Werra, A user's guide to tabu search. Ann. Oper. Res. **41**, 3–28 (May 1993)
99. S. Kirkpatrick, Optimization by simulated annealing: Quantitative studies. Journal of Statistical Physics **34**, 975–986 (1984)
100. J.H. Holland, *Adaptation in natural and artificial systems* (MIT Press, Cambridge, MA, USA, 1992)

101. T. Bäck, *Evolutionary algorithms in theory and practice: evolution strategies, evolutionary programming, genetic algorithms* (Oxford University Press, Oxford, UK, 1996)
102. M. F. Ercan and C. OÄŸuz, "Performance of local search heuristics on scheduling a class of pipelined multiprocessor tasks", Computers and Electrical Engineering, vol. 31, no. 8, pp. 537–555, 2005.
103. A. Tumeo, C. Pilato, F. Ferrandi, D. Sciuto, and P. Lanzi, "Ant colony optimization for mapping and scheduling in heterogeneous multiprocessor systems", in Embedded Computer Systems: Architectures, Modeling, and Simulation, 2008. SAMOS 2008. International Conference on, pp. 142–149, july 2008.
104. A. Tumeo, M. Branca, L. Camerini, C. Pilato, P. L. Lanzi, F. Ferrandi, and D. Sciuto, "Mapping pipelined applications onto heterogeneous embedded systems: a bayesian optimization algorithm based approach", in Proceedings of the 7th IEEE/ACM international conference on Hardware/software codesign and system synthesis, CODES+ISSS '09, (New York, NY, USA), pp. 443–452, ACM, 2009.
105. M. Branca, L. Camerini, F. Ferrandi, P. L. Lanzi, C. Pilato, D. Sciuto, and A. Tumeo, "Evolutionary algorithms for the mapping of pipelined applications onto heterogeneous embedded systems", in Proceedings of the 11th Annual conference on Genetic and evolutionary computation, GECCO '09, (New York, NY, USA), pp. 1435–1442, ACM, 2009.
106. H. Yang and S. Ha, "Pipelined data parallel task mapping/scheduling technique for mpsoc", in Design, Automation Test in Europe Conference Exhibition, 2009. DATE '09., pp. 69–74, april 2009.
107. T. Blickle, J. Teich, L. Thiele, System-level synthesis using evolutionary algorithms. Design Automation for Embedded Systems **3**(1), 23–58 (1998)
108. L. Pomante, "System-level design space exploration for dedicated heterogeneous multiprocessor systems", in Application-Specific Systems, Architectures and Processors (ASAP), 2011 IEEE International Conference on, pp. 79–86, sept. 2011.
109. C. Erbas, S. Cerav-Erbas, and A. Pimentel, "Multiobjective optimization and evolutionary algorithms for the application mapping problem in multiprocessor system-on-chip design", Evolutionary Computation, IEEE Transactions on, vol. 10, pp. 358–374, june 2006.
110. B. Meyer and D. Thomas, "Rethinking automated synthesis of mpsoc architectures", in Parallel and Distributed Processing Symposium, 2007. IPDPS 2007. IEEE, International, pp. 1–6, March 2007.
111. T. Givargis, F. Vahid, and J. Henkel, "System-level exploration for pareto-optimal configurations in parameterized system-on-a-chip", Very Large Scale Integration (VLSI) Systems, IEEE Transactions on, vol. 10, pp. 416–422, aug. 2002.
112. F. Sun, S. Ravi, A. Raghunathan, and N. K. Jha, "Synthesis of application-specific heterogeneous multiprocessor architectures using extensible processors", in VLSID '05: Proceedings of the 18th International Conference on VLSI Design held jointly with 4th International Conference on Embedded Systems Design, (Washington, DC, USA), pp. 551–556, IEEE Computer Society, 2005.
113. I. Karkowski and H. Corporaal, "Design of heterogenous multi-processor embedded systems: applying functional pipelining", in PACT '97: Proceedings of the 1997 International Conference on Parallel Architectures and Compilation Techniques, IEEE Computer Society, 1997.
114. I. Karkowski and H. Corporaal, "Design space exploration algorithm for heterogeneous multiprocessor embedded system design", in DAC '98: Proceedings of the 35th annual Design Automation Conference, (New York, NY, USA), pp. 82–87, ACM, 1998.
115. L. Chen, N. Boichat, and T. Mitra, "Customized mpsoc synthesis for task sequence", in Proceedings of the 2011 IEEE 9th Symposium on Application Specific Processors, SASP '11, (Washington, DC, USA), pp. 16–21, IEEE Computer Society, 2011.
116. U. Bordoloi, H. P. Huynh, T. Mitra, and S. Chakraborty, "Design space exploration of instruction set customizable mpsocs for multimedia applications", in Embedded Computer Systems (SAMOS), 2010 International Conference on, pp. 170–177, july 2010.

117. P. Chandraiah and R. Domer, "Code and data structure partitioning for parallel and flexible mpsoc specification using designer-controlled recoding", Computer-Aided Design of Integrated Circuits and Systems, IEEE Transactions on, vol. 27, pp. 1078–1090, june 2008.
118. J.-Y. Mignolet, R. Baert, T. Ashby, P. Avasare, H.-O. Jang, and J. C. Son, "Mpa: Parallelizing an application onto a multicore platform made easy", Micro, IEEE, vol. 29, pp. 31–39, may-june 2009.
119. G. Tournavitis, Z. Wang, B. Franke, and M. F. O'Boyle, "Towards a holistic approach to auto-parallelization: integrating profile-driven parallelism detection and machine-learning based mapping", in Proceedings of the 2009 ACM SIGPLAN conference on Programming language design and implementation, PLDI '09, (New York, NY, USA), pp. 177–187, ACM, 2009.
120. J. Ceng, J. Castrillon, W. Sheng, H. Scharwächter, R. Leupers, G. Ascheid, H. Meyr, T. Isshiki, and H. Kunieda, "Maps: an integrated framework for mpsoc application parallelization", in DAC '08: Proceedings of the 45th annual Design Automation Conference, (New York, NY, USA), pp. 754–759, ACM, 2008.
121. R. Leupers and J. Castrillon, "Mpsoc programming using the maps compiler", in Design Automation Conference (ASP-DAC), 2010 15th Asia and South Pacific, pp. 897–902, jan. 2010.
122. D. Cordes, P. Marwedel, and A. Mallik, "Automatic parallelization of embedded software using hierarchical task graphs and integer linear programming", in Hardware/Software Codesign and System Synthesis (CODES+ISSS), 2010 IEEE/ACM/IFIP International Conference on, pp. 267–276, oct. 2010.
123. D. Cordes, A. Heinig, P. Marwedel, and A. Mallik, "Automatic extraction of pipeline parallelism for embedded software using linear programming", in Parallel and Distributed Systems (ICPADS), 2011 IEEE 17th International Conference on, pp. 699–706, dec. 2011.
124. K. Huang, S.-i. Han, K. Popovici, L. Brisolara, X. Guerin, L. Li, X. Yan, S.-l. Chae, L. Carro, and A. A. Jerraya, "Simulink-based mpsoc design flow: case study of motion-jpeg and h.264", in DAC '07: Proceedings of the 44th annual Design Automation Conference, (New York, NY, USA), pp. 39–42, ACM, 2007.
125. S.-I. Han, S.-I. Chae, L. Brisolara, L. Carro, K. Popovici, X. Guerin, A.A. Jerraya, K. Huang, L. Li, X. Yan, Simulink-based heterogeneous multiprocessor soc design flow for mixed hardware/software refinement and simulation. Integration, the VLSI Journal **42**(2), 227–245 (2009)
126. D. Lyonnard, S. Yoo, A. Baghdadi, and A. A. Jerraya, "Automatic generation of application-specific architectures for heterogeneous multiprocessor system-on-chip", in DAC '01: Proceedings of the 38th annual Design Automation Conference, (New York, NY, USA), pp. 518–523, ACM, 2001.
127. P. Van Der Wolf, E. de Kock, T. Henriksson, W. Kruijtzer, and G. Essink, "Design and programming of embedded multiprocessors: an interface-centric approach", in CODES+ISSS '04: Proceedings of the 2nd IEEE/ACM/IFIP international conference on Hardware/software codesign and system synthesis, (New York, NY, USA), pp. 206–217, ACM, 2004.
128. K. Lahiri, A. Raghunathan, and S. Dey, "Design space exploration for optimizing on-chip communication architectures", Computer-Aided Design of Integrated Circuits and Systems, IEEE Transactions on, vol. 23, pp. 952–961, june 2004.
129. F. Dumitrascu, I. Bacivarov, L. Pieralisi, M. Bonaciu, and A. Jerraya, "Flexible mpsoc platform with fast interconnect exploration for optimal system performance for a specific application", in Design, Automation and Test in Europe, 2006. DATE '06. Proceedings, vol. 2, p. 6 pp., march 2006.
130. S. Pasricha and N. D. Dutt, "A framework for cosynthesis of memory and communication architectures for mpsoc", Computer-Aided Design of Integrated Circuits and Systems, IEEE Transactions on, vol. 26, pp. 408–420, march 2007.
131. A. Wieferink, T. Kogel, R. Leupers, G. Ascheid, H. Meyr, G. Braun, and A. Nohl, "A system level processor/communication co-exploration methodology for multi-processor system-on-chip platforms", in DATE '04: Proceedings of the conference on Design, automation and test in Europe, (Washington, DC, USA), p. 21256, IEEE Computer Society, 2004.

132. A. Hoffmann, T. Kogel, A. Nohl, G. Braun, O. Schliebusch, O. Wahlen, A. Wieferink, and H. Meyr, "A novel methodology for the design of application-specific instruction-set processors (asips) using a machine description language", Computer-Aided Design of Integrated Circuits and Systems, IEEE Transactions on, vol. 20, pp. 1338–1354, nov 2001.
133. F. Angiolini, J. Ceng, R. Leupers, F. Ferrari, C. Ferri, and L. Benini, "An integrated open framework for heterogeneous mpsoc design space exploration", in DATE '06: Proceedings of the conference on Design, automation and test in Europe, (3001 Leuven, Belgium, Belgium), pp. 1145–1150, European Design and Automation Association, 2006.
134. CoWare, "Lisatek". Available at: http://www.coware.com/
135. M. Loghi, F. Angiolini, D. Bertozzi, L. Benini, and R. Zafalon, "Analyzing on-chip communication in a mpsoc environment", in DATE '04: Proceedings of the conference on Design, automation and test in Europe, (Washington, DC, USA), p. 20752, IEEE Computer Society, 2004.
136. G. Martin, Multi-processor soc-based design methodologies using configurable and extensible processors. J. Signal Process. Syst. **53**(1–2), 113–127 (2008)
137. A. Gerstlauer, C. Haubelt, A. Pimentel, T. Stefanov, D. Gajski, and J. Teich, "Electronic system-level synthesis methodologies", Computer-Aided Design of Integrated Circuits and Systems, IEEE Transactions on, vol. 28, pp. 1517–1530, oct. 2009.
138. F. Balarin, Y. Watanabe, H. Hsieh, L. Lavagno, C. Passerone, and A. Sangiovanni-Vincentelli, "Metropolis: an integrated electronic system design environment", Computer, vol. 36, pp. 45–52, april 2003.
139. A. Sangiovanni-Vincentelli, "Quo vadis, sld? reasoning about the trends and challenges of system level design", Proceedings of the IEEE, vol. 95, pp. 467–506, march 2007.
140. T. Kangas, P. Kukkala, H. Orsila, E. Salminen, M. Hännikäinen, T.D. Hämäläinen, J. Riihimäki, K. Kuusilinna, Uml-based multiprocessor soc design framework. ACM Trans. Embed. Comput. Syst. **5**, 281–320 (May 2006)
141. S. Ha, S. Kim, C. Lee, Y. Yi, S. Kwon, and Y.-P. Joo, "Peace: A hardware-software codesign environment for multimedia embedded systems", ACM Trans. Des. Autom. Electron. Syst., vol. 12, pp. 24:1–24:25, May 2008.
142. University of California Berkeley, "Ptolemy project". Available at http://ptolemy.eecs.berkeley.edu/ptolemyII/
143. S. Kwon, Y. Kim, W.-C. Jeun, S. Ha, and Y. Paek, "A retargetable parallel-programming framework for mpsoc", ACM Trans. Des. Autom. Electron. Syst., vol. 13, pp. 39:1–39:18, July 2008.
144. M. Thompson, H. Nikolov, T. Stefanov, A. D. Pimentel, C. Erbas, S. Polstra, and E. F. Deprettere, "A framework for rapid system-level exploration, synthesis, and programming of multimedia mp-socs", in CODES+ISSS '07: Proceedings of the 5th IEEE/ACM international conference on Hardware/software codesign and system synthesis, (New York, NY, USA), pp. 9–14, ACM, 2007.
145. H. Nikolov, M. Thompson, T. Stefanov, A. Pimentel, S. Polstra, R. Bose, C. Zissulescu, and E. Deprettere, "Daedalus: toward composable multimedia mp-soc design", in Proceedings of the 45th annual Design Automation Conference, DAC '08, (New York, NY, USA), pp. 574–579, ACM, 2008.
146. S. Verdoolaege, H. Nikolov, T. Stefanov, pn: a tool for improved derivation of process networks. EURASIP J. Embedded Syst. **2007**, 19–19 (Jan. 2007)
147. A. Pimentel, C. Erbas, S. Polstra, A systematic approach to exploring embedded system architectures at multiple abstraction levels. Computers, IEEE Transactions on **55**(2), 99–112 (2006)
148. H. Nikolov, T. Stefanov, and E. F. Deprettere, "Systematic and automated multiprocessor system design, programming, and implementation", IEEE Trans. on CAD of Integrated Circuits and Systems, vol. 27, no. 3, pp. 542–555, 2008.
149. M. Bamakhrama, J. Zhai, H. Nikolov, and T. Stefanov, "A methodology for automated design of hard-real-time embedded streaming systems", in Design, Automation Test in Europe Conference Exhibition (DATE), 2012, pp. 941–946, march 2012.

150. R. Dömer, A. Gerstlauer, J. Peng, D. Shin, L. Cai, H. Yu, S. Abdi, and D. D. Gajski, "System-on-chip environment: a specc-based framework for heterogeneous mpsoc design", EURASIP J. Embedded Syst., vol. 2008, pp. 5:1–5:13, Jan. 2008.
151. D. Gajski, J. Zhu, R. Dömer, A. Gerstlauer, *S* (Specification Language and Design Methodology. Kluwer Academic Publishers, Zhao, SpecC, 2000)
152. J. Keinert, M. Streub&uhr, T. Schlichter, J. Falk, J. Gladigau, C. Haubelt, J. Teich, and M. Meredith, "Systemcodesigner—an automatic esl synthesis approach by design space exploration and behavioral synthesis for streaming applications", ACM Trans. Des. Autom. Electron. Syst., vol. 14, pp. 1:1–1:23, Jan. 2009.
153. M. Platzner, J. Teich, *N* (Dynamically Reconfigurable Systems. Springer, When, 2010)
154. A. Schranzhofer, J.-J. Chen, and L. Thiele, "Dynamic power-aware mapping of applications onto heterogeneous mpsoc platforms", Industrial Informatics, IEEE Transactions on, vol. 6, pp. 692–707, nov. 2010.
155. C. Ykman-Couvreur, P. Avasare, G. Mariani, G. Palermo, C. Silvano, and V. Zaccaria, "Linking run-time resource management of embedded multi-core platforms with automated design-time exploration", Computers Digital Techniques, IET, vol. 5, pp. 123–135, march 2011.
156. C. Silvano, W. Fornaciari, *E* (The Multicube approach. Springer, Villar, Multi-objective Design Space Exploration of Multiprocessor SoC Architectures, 2011)
157. K. Niyogi and D. Marculescu, "Speed and voltage selection for gals systems based on voltage/frequency islands", in Proceedings of the 2005 Asia and South Pacific Design Automation Conference, ASP-DAC '05, (New York, NY, USA), pp. 292–297, ACM, 2005.
158. C. Isci, A. Buyuktosunoglu, C.-Y. Cher, P. Bose, and M. Martonosi, "An analysis of efficient multi-core global power management policies: Maximizing performance for a given power budget", in Proceedings of the 39th Annual IEEE/ACM International Symposium on Microarchitecture, MICRO 39, (Washington, DC, USA), pp. 347–358, IEEE Computer Society, 2006.
159. D. Puschini, F. Clermidy, P. Benoit, G. Sassatelli, and L. Torres, "Dynamic and distributed frequency assignment for energy and latency constrained mp-soc", in Design, Automation Test in Europe Conference Exhibition, 2009. DATE '09., pp. 1564–1567, april 2009.
160. A. Molnos and K. Goossens, "Conservative dynamic energy management for real-time dataflow applications mapped on multiple processors", in Digital System Design, Architectures, Methods and Tools, 2009. DSD '09. 12th Euromicro Conference on, pp. 409–418, aug. 2009.
161. J. Huang, A. Raabe, C. Buckl, and A. Knoll, "A workflow for runtime adaptive task allocation on heterogeneous mpsocs", in Design, Automation Test in Europe Conference Exhibition (DATE), 2011, pp. 1–6, march 2011.
162. P. Malani, P. Mukre, Q. Qiu, and Q. Wu, "Adaptive scheduling and voltage scaling for multiprocessor real-time applications with non-deterministic workload", in Proceedings of the conference on Design, automation and test in Europe, DATE '08, (New York, NY, USA), pp. 652–657, ACM, 2008.
163. K. Huang, L. Santinelli, J.-J. Chen, L. Thiele, and G. Buttazzo, "Adaptive dynamic power management for hard real-time systems", in Real-Time Systems Symposium, 2009, RTSS 2009. 30th IEEE, pp. 23–32, dec. 2009.
164. A. Coskun, T. Rosing, and K. Gross, "Utilizing predictors for efficient thermal management in multiprocessor socs", Computer-Aided Design of Integrated Circuits and Systems, IEEE Transactions on, vol. 28, pp. 1503–1516, oct. 2009.
165. T. Ebi, M. Faruque, and J. Henkel, "Tape: Thermal-aware agent-based power econom multi/many-core architectures", in Computer-Aided Design - Digest of Technical Papers, 2009. ICCAD 2009. IEEE/ACM International Conference on, pp. 302–309, nov. 2009.
166. Y. Ge and Q. Qiu, "Dynamic thermal management for multimedia applications using machine learning", in Proceedings of the 48th Design Automation Conference, DAC '11, (New York, NY, USA), pp. 95–100, ACM, 2011.
167. Y. Tan, W. Liu, and Q. Qiu, "Adaptive power management using reinforcement learning", in Proceedings of the 2009 International Conference on Computer-Aided Design, ICCAD '09, (New York, NY, USA), pp. 461–467, ACM, 2009.

168. E. de Souza Carvalho, N. Calazans, and F. Moraes, "Dynamic task mapping for mpsocs", Design Test of Computers, IEEE, vol. 27, pp. 26–35, sept.-oct. 2010.
169. M. A. Al Faruque, R. Krist, and J. Henkel, "Adam: run-time agent-based distributed application mapping for on-chip communication", in Proceedings of the 45th annual Design Automation Conference, DAC '08, (New York, NY, USA), pp. 760–765, ACM, 2008.

Chapter 3
Optimisation Framework

This chapter provides a philosophical overview of the research reported in this monograph. Firstly, the application model and pipelined MPSoCs considered in this monograph are described. Then, shortcomings of prior research on pipelined MPSoCs are discussed in order to provide an idea of how this monograph fills in some of the gaps in prior research. Lastly, this chapter rationalises the design-time and run-time optimisations proposed for pipelined MPSoCs in this monograph.

3.1 Application Model and Pipelined MPSoCs

Multimedia applications are characterised by several sub-kernels which are executed repeatedly on an input data stream. For example, the JPEG decoder application contains the following sub-kernels: Entropy Decoding (ED); Inverse Transformation and Quantisation (ITQ); and, Colour Conversion (CC). These sub-kernels are independent of each other and hence can operate on different data units at the same time, enabling their execution on pipelined MPSoCs. The number of invocations of all the sub-kernels is the same and equal to the number of data units in the input. Hence, the number of iterations of the application is equal to the number of times each sub-kernel is invoked. Figure 3.1 illustrates task graphs of several multimedia applications where nodes and edges represent the sub-kernels and the data dependencies respectively.

In a pipelined MPSoC, processors are organised in stages where the stages are connected in a pipeline. Communication between the stages typically occurs through FIFO buffers. These FIFO buffers allow communication at a much higher bandwidth compared to a shared bus and provide blocking read and write operations to allow synchronisation between the processors running at different frequencies. Each processor is an Application Specific Instruction set Processor (ASIP) with separate instruction and data caches that are connected to its local memory. In addition to local memories, shared memory could be used where common data need to be shared within a stage

H. Javaid and S. Parameswaran, *Pipelined Multiprocessor System-on-Chip for Multimedia*, DOI: 10.1007/978-3-319-01113-4_3,

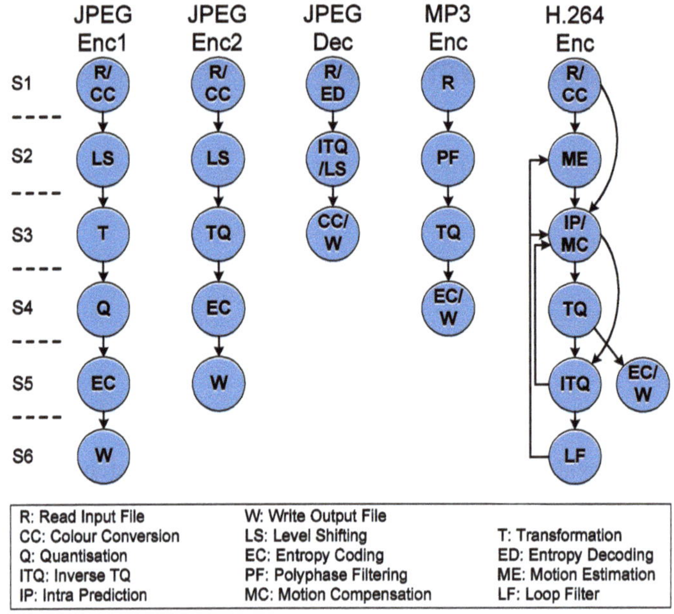

Fig. 3.1 Graphs of typical multimedia applications

and/or among different stages. The nodes and edges of a multimedia application's task graph are assigned to one or more processors and FIFO buffers in the pipelined MPSoC. For example, the sub-kernels of JPEGEnc1 have been one-to-one mapped onto a pipelined MPSoC shown in Fig. 3.2. While the processor P2.1 is in *i*th iteration, P1.1 will be in its (*i*+1)th iteration, thereby allowing pipelined execution of the sub-kernels. In other words, the input data streams through the pipelined MPSoC before being written by the last stage. Note that an iteration of the pipelined MPSoC refers to the processing of one data unit by all sub-kernels, and the number of iterations of a pipelined MPSoC is the number of iterations of the application executing on it. Figure 3.2 also shows pipelined MPSoCs for other applications where some of the FIFO buffers and memories have been omitted for the sake of simplicity. Note that the TQ sub-kernel of JPEGEnc2 is implemented on three processors, which will work in parallel to achieve the required performance.

The backward edges in an application graph introduce data dependencies that can hamper pipelined execution of the sub-kernels. This is because the stage which requires data from a backward edge (consumer stage) might have to wait due to a stage further in the pipeline (producer stage). For example, stage 2 (ME) of the H.264Enc requires data from stage 6 (LF) in Fig. 3.1. The *dependency distance* of an edge is the dependence distance in number of iterations between the consumer and the producer stages. Consider that the *i*th iteration of stage 2 needs the output of (*i*-7)th iteration of stage 6, then the dependency distance of the backward edge is seven. To avoid unnecessary waiting due to a backward edge in a balanced pipelined

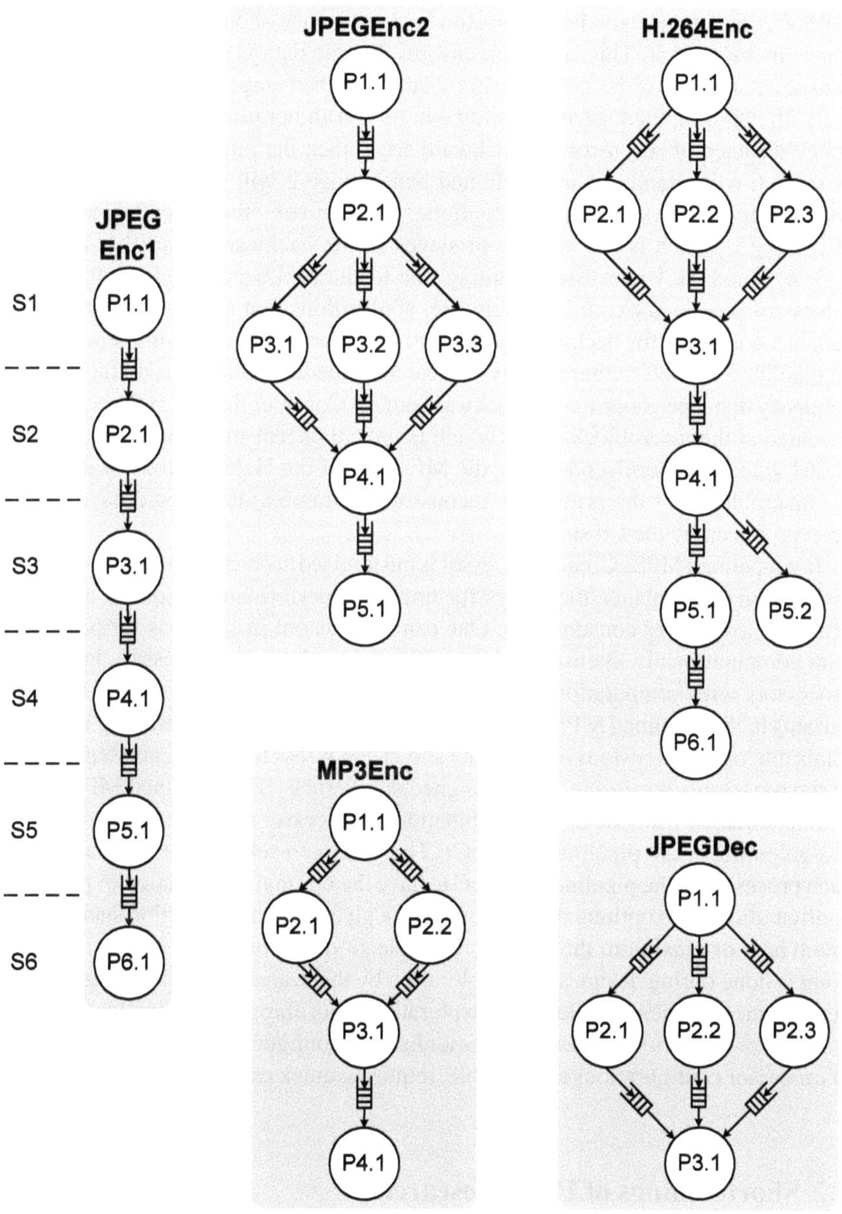

Fig. 3.2 Pipelined MPSoCs for multimedia applications of Fig. 3.1 (some FIFO buffers and memories are not shown for the sake of simplicity)

MPSoC, the dependency distance of the backward edge should be $\geq$ the number of stages included in it. This condition ensures that the data is always available to the consumer stage on or before time. For example, when stage 2 of H.264 encoder is in its ith iteration, then stage 6 would be in its (i-4)th iteration. Consider the dependency distance of seven for the backward edge, then the output of (i-7)th iteration of stage 6 will already be available and hence stage 2 will not wait unnecessarily and the pipelined execution will continue. Alternatively, the dependency distance of 7 is $\geq$ 5, which is the number of stages in the backward edge (that is, stages 2, 3, 4, 5 and 6). Unnecessary waiting due to the backward edges voids the usefulness of pipelined execution; therefore, applications that violate the dependency distance condition for backward edges will not benefit from their implementations on pipelined MPSoCs. Interestingly, typical multimedia applications do fulfil the dependency distance condition for backward edges. Consider that the H.264 encoder is executed at the macroblock-level (which is typical of real-time implementations of H.264 video encoder/decoder [1]), the ME stage of the H.264 encoder will require the macroblocks of the previously reconstructed frame(s) that would have already been produced by the LF stage.

In a pipelined MPSoC, each processor is customised according to the sub-kernel(s) assigned to it to balance the stages for improved performance, reduced area footprint and low power consumption. One can add custom instructions for processors with computationally intensive sub-kernels while reducing unnecessary logic from processors with computationally light sub-kernels. Hence, at the system-level, the variants in the pipelined MPSoC are the processor configurations resulting from customisable options (custom instructions and cache sizes) that are generated for each of the processors according to the assigned sub-kernels. The pipelined MPSoC will be implemented with one of the combination of processor configurations (one of the design points of the pipelined MPSoC). The goal is to select one configuration for each processor in the pipelined MPSoC to have the optimal combination of processor configurations (the optimal design point) for a given objective function such as minimum area or maximum throughput. The selection of a pipelined MPSoC's design point is done during design space exploration by the evaluation of the design points' performance metrics, coupled with exploration algorithms. For a pipelined MPSoC with five processors where each processor has 100 configurations, 10^{10} combinations of processor configurations are possible, requiring quick exploration methodologies.

3.2 Shortcomings of Prior Research

To enable quick exploration of a pipelined MPSoC's design space, a quick methodology to obtain performance metrics of all the design points is required. Since there can be billions of design points, a simulation only methodology is not feasible. A few works on performance estimation of pipelined MPSoCs used full—system, cycle-accurate simulations and an analytical model for only the execution time [2–5]. The works in [2, 3] proposed less accurate models, while [4, 5] did not evaluate the accuracy of their models. Chapter 4 addresses these issues by introducing analytical models for three performance metrics (execution time, latency and throughput) of a pipelined MPSoC and evaluates their absolute accuracy and fidelity [6]. Throughput and latency are typical performance metrics for real-time pipelined MPSoCs. Two estimation methods are also proposed in Chap. 4 to reduce the number of full-system, cycle-accurate simulations of the pipelined MPSoC to aid quick exploration.

Once the performance metrics of design points are available (or can be computed quickly), the next step is to use exploration algorithms to search for the optimal design point. Jin et al. [7] addressed the problem of maximising the throughput of a multimedia application on a pipelined MPSoC with fixed number of processors. Cong et al. [8] proposed exact algorithms to minimise latency and number of processors in a pipelined MPSoC under a throughput constraint. Both these works [7, 8] did not consider processor customisation, and thus dealt with homogeneous pipelined MPSoCs only.

The work in [2] addressed the problem of processor customisation (selection of custom instructions or selection of processor configurations) in a pipelined MPSoC. Shee et al. [2] proposed a heuristic to maximise pipelined MPSoC's execution time improvement per area increase ratio compared to a uniprocessor system. Thus, Shee et al. did not consider performance constraints that are typical of real-time multimedia applications. Chapter 5 addresses these issues by proposing three techniques for area footprint optimisation of a pipelined MPSoC under an execution time or a latency or a throughput constraint respectively. To speed up the exploration process, these techniques utilise the performance analytical models and estimation methods proposed in Chap. 4. Two works inspired from the proposals of Chap. 5 have been published recently [9, 10]. Bordoli et al. [9] considered variations in processor latencies during customisation of the processors. Their objective was to minimise variation in throughput under an area footprint constraint. Chen et al. [10] explored simultaneous mapping and processor customisation with variable number of processors in the pipelined MPSoC. Their aim was to minimise MPSoC's area under a throughput constraint.

The pipelined MPSoCs optimised at design-time use worst-case parameters to ensure that the performance required of them is delivered at all times when deployed. As such, worst-case pipelined MPSoCs lack run-time adaptability, and thus may result in inefficient resource utilisation and increased energy consumption under a dynamic workload. Hence, to enable low-power operation under a dynamic workload, run-time adaptability must be introduced in pipelined MPSoCs.

The works in [11–13] considered run-time adaptability in pipelined MPSoCs. Guo et al. [13] proposed a dynamic voltage scaling approach to reduce the voltage to processors with low workload, while [11, 12] showed the application of Dynamic Voltage and Frequency Scaling (DVFS) in pipelined MPSoCs. All these works used a feedback controller to monitor the occupancy level of the FIFO buffers to determine when to increase or decrease the frequency-voltage levels of a processor. Although feedback controllers provide run-time adaptability in pipelined MPSoCs, they are reactive in nature rather than proactive as they do not utilise any form of prediction. Therefore, the controller only acts when a performance penalty has occurred instead of forecasting such a penalty and acting in advance. Proactive techniques are typically required when the variations in workload are sudden due to input data dependence, which is the case with multimedia applications. Chapters 6 and 7 address run-time adaptability issues in pipelined MPSoCs. Chapter 6 introduces an adaptive pipelined MPSoC architecture with a processor manager to predict idle processors in the pipelined MPSoC at run-time. The processor manager not only utilises the application's execution history, but also the application's knowledge to predict the upcoming workload. An application's knowledge should be used in workload prediction because an application knows (or may know) by far the most about its future workload [14]. The idle processors are either clock- or power-gated to illustrate the energy efficiency of adaptive pipelined MPSoCs compared to worst-case pipelined MPSoCs. Thus, Chap. 6 proposes proactive rather than reactive run-time processor management techniques for adaptive pipelined MPSoCs.

Practically, provision of the DVFS circuitry for MPSoCs with more than two processors is very expensive [15]. Furthermore, the large overhead of the DVFS control circuitry limits its use to systems requiring only coarse-grained run-time management [16]. The shrinkage of the dynamic range of frequency-voltage operational points due to downward scaling of supply voltage has also limited the use of DVFS, and has given rise to the use of clock-gating, power-gating and multiple power states. Therefore, Chap. 6 used either clock- or power-gating to deactivate idle processors in an adaptive pipelined MPSoC. Chapter 7 extends this work for multiple power states, where the challenge is to also predict the upcoming idle period of an idle processor to select the most energy saving power state. Like Chaps. 6, 7 also proposes proactive run-time techniques utilising the application's knowledge, but for power management of adaptive pipelined MPSoCs.

A pipelined MPSoC will typically be used as a multimedia accelerator in a multimedia platform (such as OMAP [17], Tegra [18], etc.) because it is extremely customised for a specific multimedia application. So far, both worst-case (non-adaptive) and adaptive pipelined MPSoCs have been designed for only one multimedia application, requiring the deployment of individual accelerators for multimedia applications. Due to the area constraints in portable media devices, it is desirable to use a multi-mode accelerator rather than individual accelerators when their use is mutually exclusive. Chapter 8 makes the first attempt at multi-mode pipelined MPSoCs for multiple, mutually exclusive applications to function as multi-mode multimedia accelerators where each mode refers to the execution of one application. The idea of merging individual application graphs into a single application graph at design-time is exploited for realisation of a multi-mode pipelined MPSoC.

3.3 Overview of Optimisation Framework

The aim of this monograph is to optimise pipelined MPSoCs by reducing their area footprint and lowering their power consumption under real-time performance constraints. The author proposes design-time and run-time optimisations, which are targeted at different objective functions. At first, a pipelined MPSoC is optimised for area footprint under an execution time, a latency constraint or a throughput constraint. Then, such a design-time optimised pipelined MPSoC is augmented with run-time adaptability for low-power operation under a dynamic workload. Finally, the pipelined MPSoCs optimised for different multimedia applications are combined into a single multi-mode pipelined MPSoC for further reduction of the area footprint. Figure 3.3 presents a philosophical overview of how the research reported in different chapters of this monograph is connected, and can be used to optimise pipelined MPSoCs in the form of an optimisation framework.

The first phase of the framework involves analysis and profiling of the multimedia application to decide the initial architecture of the pipelined MPSoC (number of processors, and number, size and connection of the FIFO buffers). The analysis involves extraction of the sub-kernels if the application is specified as a C/C++ code, which can be done manually or semi-automatically [19, 20]. Alternatively, an application can be specified in a stream language such as StreamIt [21] to explicitly represent the sub-kernels. Once the sub-kernels are available, profiling is performed to analyse the computational ratios of the sub-kernels so that they can be merged and/or split if required. This process is typically referred to as an application-level balancing [22–24] and is done to ensure that the sub-kernels contain reasonable amount of computation to be mapped to individual processors. After such an application-level balancing, code segments of the sub-kernels are produced, in addition to the application graph where nodes represent sub-kernels and edges represent data dependencies. The initial pipelined MPSoC is then derived by mapping sub-kernels and edges of the application graph to one or more processors and FIFO buffers respectively. Note that the first phase of the framework is done manually or semi-automatically by the designer, and is not the focus of this monograph.

The second phase of the framework optimises the area footprint of the initial pipelined MPSoC under a performance constraint by customising the processors. This phase takes the code segments of the sub-kernels and the pipelined MPSoC architecture as the input from the last phase (the application graph is used in the fourth phase). Initially, for each of the processors in the pipelined MPSoC, processor configurations trading-off performance and area footprint are created by combining the custom instructions (generated using an ASIP generator which analyses the code segments of the sub-kernels) with cache configurations. The combinations of these processor configurations make up the design points of the pipelined MPSoC's design space. The goal is to quickly explore the design space (by utilising quick performance evaluation of design points and fast exploration algorithms) to select one configuration for each processor so that the area of the pipelined MPSoC is minimised under a performance constraint.

Chapter 4 proposes analytical models to estimate the execution time, latency and throughput of a pipelined MPSoC's design point using latencies of individual processor configurations, and thus avoiding slow, full-system, cycle accurate simulations of all the design points. For effective use of these analytical models, latencies of individual processor configurations should be available. To this end, two estimation methods (PS and PSP) are proposed to gather latencies of processor configurations with the minimal number of simulations. The PS method simulates all the processor

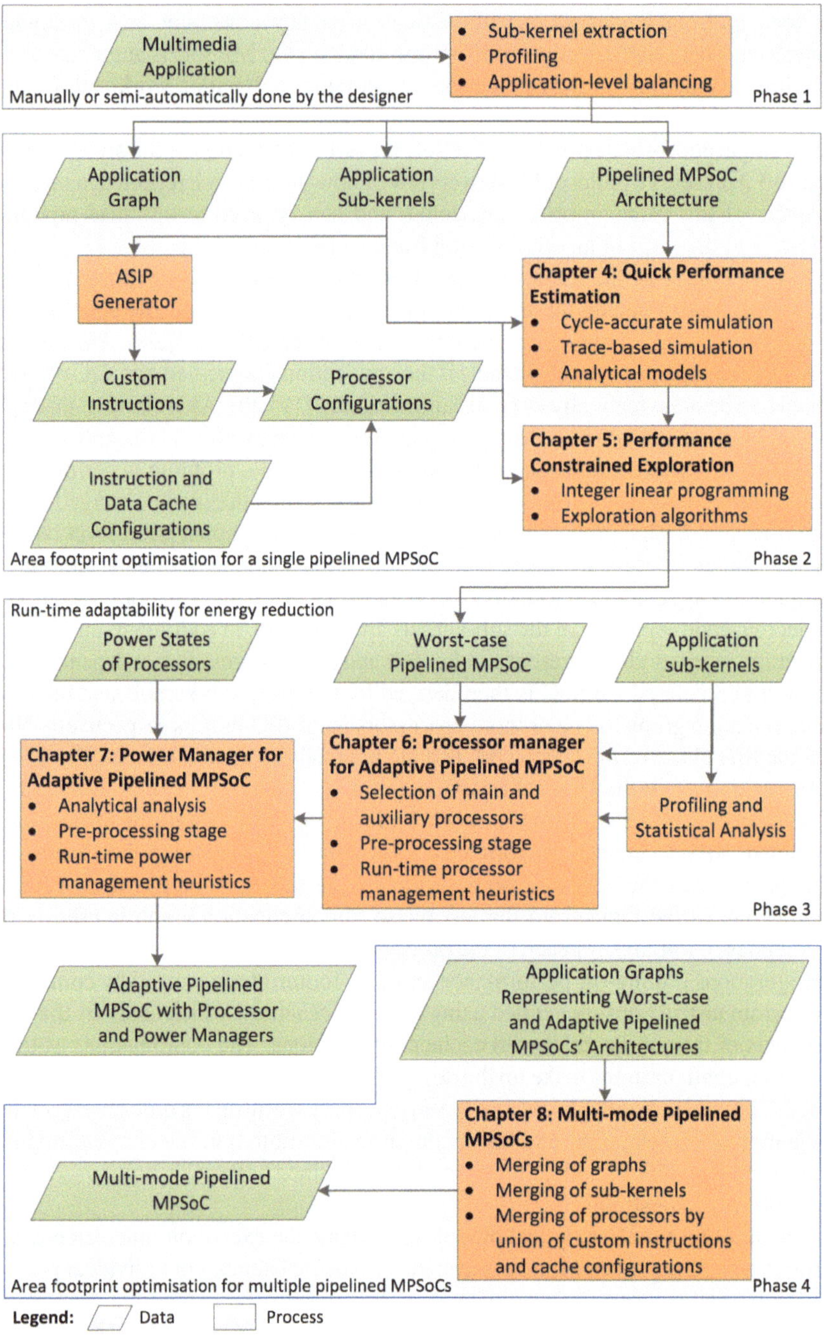

Fig. 3.3 An optimisation framework for pipelined MPSoCs

configurations once, while the PSP method simulates only a subset of processor configurations and then uses a processor analytical model to estimate the latencies of the processor configurations. Experiments with a number of pipelined MPSoCs executing typical multimedia applications (JPEG encoder/decoder, MP3 encoder and H.264 encoder) showed that the analytical models with PS and PSP methods had maximum absolute errors of 12.95 and 18.67 % respectively, and minimum fidelities of 0.93 and 0.88 respectively. The design spaces of the pipelined MPSoCs ranged from 10^{12} to 10^{18} design points, and hence simulation of all design points will take years and is infeasible. Compared to the PS method, the PSP method reduced simulation time from days to several hours because it reduced the number of simulations from hundreds to only tens.

Chapter 5 builds upon Chap. 4 by utilising the analytical models in the exploration algorithms for quick design space exploration. It proposes Integer Linear Programming (ILP) based techniques for area footprint optimisation under an execution time constraint or a latency constraint, and an algorithm for area footprint optimisation under a throughput constraint. The proposed exploration techniques were evaluated on the five pipelined MPSoCs created in Chap. 4, again with design spaces up to 10^{18} design points. The time to find the Pareto front of each pipelined MPSoC with respect to execution time or latency or throughput was less than seven minutes, illustrating the applicability of the proposed design space exploration method. At the end of the second phase, implementation of the pipelined MPSoC in terms of the processor configurations is available. Note that a designer will typically optimise the pipelined MPSoC using worst-case latencies of the processor configurations (that is, processor latencies will be gathered by providing worst-case representative input data to the pipelined MPSoCs) so that it can deliver the throughput required of it at all times when deployed.

Once a pipelined MPSoC optimised for area footprint is available, the third phase of the framework optimises it for low power consumption with the addition of run-time adaptability. This phase takes the pipelined MPSoC optimised for area footprint using worst-case parameters and the application sub-kernels from the last phases. Initially, off-line profiling and statistical analysis of the application sub-kernels is conducted by executing the worst-case pipelined MPSoC with differing input data to record possible run-time workload variations such as average workload, standard deviation of the workload, etc. The author then exploits the fact that all the processors will not be active at all times due to dynamic workload and hence can be managed at run-time to reduce energy consumption.

Chapter 6 proposes an adaptive pipelined MPSoC architecture, capable of adapting itself to run-time variations in workload. In an adaptive pipelined MPSoC, stages with significant run-time variations in workload are implemented using *Main Processors* and *Auxiliary Processors*, where the main processor uses differing number of auxiliary processors, considering run-time workload variations. A main processor is equipped with a run-time processor management technique which uses a combination of the application's execution and knowledge (algorithmic and data properties) and information from off-line profiling and statistical analysis to proactively predict the number of auxiliary processors that should be used. The idle auxiliary processors are either clock- or power-gated to reduce energy consumption. Experiments with an H.264 video encoder, designed for HD720p at 30 fps, showed that an adaptive pipelined MPSoC provided an energy reduction of up to 34 and 39 % for clock- and power-gating based deactivation of auxiliary processors respectively with a minimum throughput of 28.75 fps compared to a worst-case pipelined MPSoC.

Chapter 7 builds upon the processor manager of Chap. 6 by proposing a power manager where auxiliary processors have multiple power states, trading-off overhead of the transition to power states with their possible energy reductions. In the presence of multiple low-power states, the challenge is to predict the duration of the idle period so that the most beneficial power state can be selected for an idle auxiliary processor. Five heuristics are proposed as part of the power manager to forecast at run-time the duration of upcoming idle period of an auxiliary processor using either the application's execution history or the application's knowledge. Then, based on the predicted duration of the idle period, the most suitable power state is selected. Compared to the use of processor manager with only clock-gating or only power-gating in an adaptive pipelined MPSoC executing H.264 video encoder (HD720p at 30 fps), the power manager reduced up to a further 40 % energy consumption with only an additional 0.5 % degradation of the throughput.

The second and third phases of the framework optimise a single pipelined MPSoC for the area footprint and energy consumption. To further reduce area footprint, processors and FIFO buffers of multiple pipelined MPSoCs, designed for multiple multimedia applications, are shared when their use is mutually exclusive by creating a multi-mode pipelined MPSoC. The third phase uses the application graphs as the representation of the pipelined MPSoCs' architectures (number of processors, and number, size and connection of the FIFO buffers). Chapter 8 proposes to merge application graphs into a single graph by finding a maximal overlap between the graphs so that the multi-mode pipelined MPSoC derived from the merged graph contains minimal resources. The results indicate significant area footprint reduction (up to 62 % processor area, 57 % FIFO area and 44 processor/FIFO ports) with minuscule degradation of system throughput (up to 2 %) and latency (up to 2 %), and an increase in energy per iteration (up to 3 %) when compared to individual pipelined MPSoCs. Chapter 8 makes the first attempt at multi-mode pipelined MPSoCs.

3.4 Summary

This chapter pointed out shortcomings of the prior research done on pipelined MPSoCs, and then explained how the research reported in this monograph addresses some of those shortcomings. The chapter then introduced an optimisation framework, where the connection between the research reported in the rest of the chapters of this monograph was illustrated.

References

1. T.C. Chen, C.J. Lian, L.G. Chen, Hardware architecture design of an h.264/avc video Codec, in *Proceedings of the 2006 Asia and South Pacific Design Automation Conference (ASP-DAC '06)*, (IEEE Press, 2006)
2. S.L. Shee, S. Parameswaran, Design methodology for pipelined heterogeneous multiprocessor system, in *Proceedings of the 44th Annual Conference on Design Automation (DAC '07)* (2007), pp. 811–816
3. I. Karkowski, H. Corporaal, Design of heterogenous multi-processor embedded systems: applying functional pipelining, in *Proceedings of the 1997 International Conference on Parallel Architectures and Compilation Techniques (PACT '97)* (IEEE Computer Society, 1997)
4. H. Javaid, S. Parameswaran, Synthesis of heterogeneous pipelined multiprocessor systems using ilp: jpeg case study, in *Proceedings of the 6th IEEE/ACM/IFIP International Conference on Hardware/Software Codesign and System Synthesis (CODES/ISSS '08)* (ACM, New York, 2008), pp. 1–6
5. H. Javaid, S. Parameswaran, A design flow for application specific heterogeneous pipelined multiprocessor systems, in *Proceedings of the 46th Annual Design Automation Conference (DAC '09)* (ACM, New York, 2009), pp. 250–253
6. H. Javaid, A. Ignjatovic, S. Parameswaran, Fidelity metrics for estimation models, in *Proceedings of the 2010 IEEE/ACM International Conference on Computer-Aided Design (ICCAD)*, Nov 2010, pp. 1–8
7. Y. Jin, N. Satish, K. Ravindran, K. Keutzer, An automated exploration framework for fpga-based soft multiprocessor systems, in *Proceedings of the 3rd IEEE/ACM/IFIP International Conference on Hardware/software Codesign and System Synthesis (CODES+ISSS '05)* (ACM, New York, 2005), pp. 273–278
8. J. Cong, G. Han, W. Jiang, Synthesis of an application-specific soft multiprocessor system, in *Proceedings of the 2007 ACM/SIGDA 15th International Symposium on Field Programmable Gate Arrays (FPGA '07)* (ACM, New York, 2007), pp. 99–107
9. U. Bordoloi, H.P. Huynh, T. Mitra, S. Chakraborty, Design space exploration of instruction set customizable MPSoCS for multimedia applications, in *Proceedings of the 2010 International Conference on Embedded Computer Systems (SAMOS)*, July 2010, pp. 170–177
10. L. Chen, N. Boichat, T. Mitra, Customized MPSoC synthesis for task sequence, in *Proceedings of the 2011 IEEE 9th Symposium on Application Specific Processors (SASP '11)* (IEEE Computer Society, Washington DC, 2011), pp. 16–21
11. S. Carta, A. Alimonda, A. Pisano, A. Acquaviva, L. Benini, A control theoretic approach to energy-efficient pipelined computation in MPSoCS. ACM Trans. Embed. Comput. Syst. **6**(4) (2007)
12. A. Alimonda, S. Carta, A. Acquaviva, A. Pisano, L. Benini, A Feedback-based approach to DVFS in data-flow applications. IEEE Trans. CAD Integr. Circ. Syst. **28**(11), 1691–1704 (2009)
13. H. Guo, S. Parameswaran, Balancing system level pipelines with stage voltage scaling, in *Proceedings of the IEEE Computer Society Annual Symposium on VLSI (ISVLSI '05)* (New Frontiers in VLSI Design, 2005)
14. X. Liu, P.J. Shenoy, M.D. Corner, Chameleon: application-level power management. IEEE Trans. Mob. Comput. **7**(8), 995–1010 (2008)
15. W. Kim, M. Gupta, G.-Y. Wei, D. Brooks, System level analysis of fast, per-core dvfs using on-chip switching regulators, in *Proceedings of the IEEE 14th International Symposium on High Performance Computer Architecture (HPCA 2008)* (2008), pp. 123–134
16. K.K. Rangan, G. Yeon Wei, D. Brooks, Thread motion: fine-grained power management for multi-core systems, in *Proceedings of the International Symposium on Computer Architecture* (2009), pp. 302–313
17. Texas instruments, Omap mobile processors, http://www.ti.com/
18. NVIDIA, Tegra multiprocessor architecture, http://www.nvidia.com/
19. S. Verdoolaege, H. Nikolov, T. Stefanov, pn: a tool for improved derivation of process networks. EURASIP J. Embed. Syst. **2007**, 19 (2007)

20. D. Cordes, A. Heinig, P. Marwedel, A. Mallik, Automatic extraction of pipeline parallelism for embedded software using linear programming, in *Proceedings of the 2011 IEEE 17th International Conference on Parallel and Distributed Systems (ICPADS)*, Dec 2011, pp. 699–706
21. W. Thies, M. Karczmarek, S.P. Amarasinghe, Streamit: a language for streaming applications, in *Proceedings of the 11th International Conference on Compiler Construction (CC '02)* (Springer, Heidelberg, 2002), pp. 179–196
22. M. Kudlur, S. Mahlke, Orchestrating the execution of stream programs on multicore platforms, in *Proceedings of the ACM SIGPLAN Conference on Programming Language Design and Implementation (PLDI ' 08)* (2008)
23. M. Hashemi, S. Ghiasi, Throughput-driven synthesis of embedded software for pipelined execution on multicore architectures. ACM Trans. Embed. Comput. Syst. **8**, 11:1–11:35 (2009)
24. S.M. Farhad, Y. Ko, B. Burgstaller, B. Scholz, Orchestration by approximation: mapping stream programs onto multicore architectures, in *Proceedings of the Sixteenth International Conference on Architectural Support for Programming Languages and Operating Systems (AS-PLOS '11)* (2011)

Chapter 4
Performance Estimation of Pipelined MPSoCs

Estimation through analytical models enables quick evaluation of design points and hence speeds up the design space exploration process. It is particularly so in the design of MPSoCs where large design spaces arise from the presence of various architectural parameters such as processor types, cache sizes and hardware accelerators. This chapter focuses on analytical models and estimation methods for three performance metrics (execution time, latency and throughput) of pipelined MPSoCs to speed up their design space exploration process. The variants in the pipelined MPSoC are the processor configurations resulting from customizable options (custom instructions and cache sizes). The selection of a combination of processor configurations (pipelined MPSoC's design point) is done during design space exploration through the evaluation of the design points' performance metrics, which are typically obtained through full-system, cycle-accurate simulations. Since there can be billions of design points, a simulation only methodology is not feasible.

The analytical models proposed in this chapter use latencies of individual processor configurations to estimate the performance of a pipelined MPSoC's design point, and hence avoid the use of slow, full-system, cycle-accurate simulations for all the design points. The analytical models are further augmented with two estimation methods (PS and PSP) to gather latencies of processor configurations with minimal number of simulations. The PS method simulates all the processor configurations once. On the other hand, the PSP method simulates a subset of processor configurations and then uses an analytical model of the processor to estimate latencies of processor configurations.

Prior research on pipelined MPSoCs' performance estimation used full-system, cycle-accurate simulations and an analytical model for only the execution time of the pipelined MPSoC [1–4]. The works in [1, 2] proposed less accurate models (see Sect. 4.4.4), while [3, 4] did not evaluate the accuracy of their models. In contrast, this chapter introduces analytical models for three performance metrics (execution time, latency and throughput) and evaluates their absolute accuracy and fidelity. Furthermore, two estimation methods are proposed to reduce the number of full-system, cycle-accurate simulations of the pipelined MPSoC.

H. Javaid and S. Parameswaran, *Pipelined Multiprocessor System-on-Chip for Multimedia*, DOI: 10.1007/978-3-319-01113-4_4,

Performance estimation of processors is typically done either through processor simulation or processor modelling. In the simulation domain, cycle-accurate simulators such as Xtensa Instruction Set Simulator (ISS) [5], PTLSim [6], RealView ARMulator ISS [7], etc. are available for various architectures. However, such simulators are slow and produce large amounts of output, and hence are not suitable for exploration of billions of design points. Processor modelling involves analytical models to capture a processor's micro-architecture and cache hierarchy timing to estimate the execution time of the application executing on it. Analytical models are less expensive to run compared to cycle-accurate simulators; however, they trade-off simulation speed with accuracy. The authors of [8] proposed a MonteCarlo based model for predicting the performance of Itanium-2 processor. The model broke down the execution time of a processor in net time to execute instructions and the stalls due to data dependencies and cache misses.

Processor configurations typically differ by the additional custom instructions and special hardware units (Instruction Set Architecture (ISA)), and the size, line size and associativity of instruction and data caches (cache configuration). A processor configuration is then a combination of an ISA and a cache configuration. Typically, there are far more cache configurations than the ISAs [1, 9]. Trace-based cache simulation [10–12] is an attractive alternative to cycle-accurate simulation of all the processor configurations. Trace-based cache simulation captures cache hit and miss statistics of all the cache configurations which are then used with an analytical model to estimate a processor configuration's execution time. Although a fast method, cache statistics do not contain sufficient timing information for absolutely accurate estimation. Singleton et al. [13] exploited cache statistics to estimate the execution time of the tasks executing on a processor. These estimated values were used in Dynamic Voltage and Frequency Scaling (DVFS) techniques to reduce the energy consumption of the processor. In contrast, the author uses cache statistics to estimate the latencies of sub-kernels on different processor configurations in a pipelined MPSoC.

Lee et al. [14] and Joseph et al. [15] proposed a linear regression based model using a wide range of predictors to estimate the execution time and power consumption of a processor. The authors of [16] modeled an out-of-order superscalar processor at a very detailed micro-architectural level by considering the effects on the Clock cycle Per Instruction (CPI) of the ISA, branch miss-prediction, the commit and reorder buffer, and instruction and data cache misses. The works in [14–16] are orthogonal to the author's processor analytical model proposed in this chapter, thus their proposals can be used to further refine and improve the proposed model at the cost of more complex analysis of the processor micro-architecture. The processor analytical model proposed in this chapter is targeted at maximally reducing the number of simulations due to cache configurations for a given ISA. Hence, the aim is to gather latencies of processor configurations with reasonable accuracy without the need for a complex, highly accurate model that will slow down the exploration of billions of design points.

4.1 Pipelined MPSoC's Analytical Models

The execution time of a pipelined MPSoC is defined as the total time taken by the multimedia application to process all the input data units. The latency of a pipelined MPSoC is the time taken to process one data unit during the steady state, which equals the time interval between the reading of the data unit by the first stage and the corresponding output by the last stage. Throughput of a pipelined MPSoC, on the other hand, is the number of data units produced per unit time by the last stage during steady state. The notion of "steady state" of a pipelined MPSoC excludes the time required to fill the pipelined MPSoC, that is, the time until the first output of the pipelined MPSoC.

Figure 4.1a shows a pipelined MPSoC with three stages, where each stage contains one processor. Each processor is annotated with a 3-tuple number, depicting the number of iterations of the sub-kernel executing on it, the *computation* latency of each iteration, and the number of words transferred in each iteration. For example, (7, 250, 64) means the sub-kernel is executed seven times, while computation latency of each iteration is 250 clock cycles and 64 words are transferred in each iteration. Since the last processor is writing out to file, 0 words will be transferred. Consider that there are no stalls between the processors, and a word transfer takes a single clock cycle, then the latency of the first processor for each iteration will be $250 + 64 = 314$ clock cycles. Likewise, the latency of the second and third processors will be 878 and 564 clock cycles respectively, which are marked in Fig. 4.1a.

A processor is considered critical in a pipelined MPSoC if its latency is the maximum from amongst all the processors. In the running example, processor 2 is the

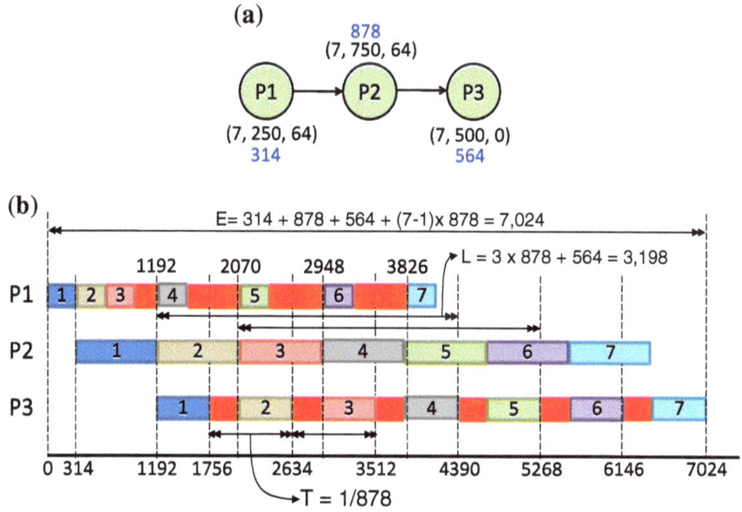

Fig. 4.1 Execution of a pipelined MPSoC

critical processor and will be the bottleneck of the system. Here, the author assumes that the intermediate FIFO buffers are able to accommodate the output of at least one iteration. For example, 64 words are transferred between processor 1 and 2, and thus the size of the FIFO buffer should be at least 64 words. Otherwise, processor 2 (which is the critical processor) will be stalled due to the limited data space in the FIFO buffer. The critical processor should not be stalled due to non-critical processors because such stalling will compromise the performance of the pipelined MPSoC. Thus, the author believes that it is reasonable to assume the availability of sufficiently-sized FIFO buffers, and hence, in the rest of the monograph, it is assumed that each FIFO buffer is able to accommodate the output of one iteration.

Figure 4.1b illustrates the execution of the pipelined MPSoC shown in Fig. 4.1a for 7 iterations. The first iteration of each processor corresponds to the filling of the pipeline. In this example, output from the first iteration of processor 1 will be available after 314 clock cycles, followed by the first output of the second processor at 1,192 clock cycles. While processor 2 is in its first iteration, processor 1 can finish its second and third iterations. However, the output of the first processor's third iteration cannot be written to the FIFO buffer, because the buffer is still holding the output of the second iteration. Thus, the first processor is stalled until the second processor reads from the FIFO buffer (until 1,192 clock cycles), which is marked with a red-coloured, unnumbered rectangle in Fig. 4.1b. At 1,192 clock cycles, processor 3 starts its first iteration, and the first output from the pipelined MPSoC is available at 1,756 clock cycles. At the same time as the start of processor 3's first iteration, that is, at 1,192 clock cycles, processor 2 will start its second iteration, emptying the FIFO buffer between processor 1 and 2. Thus, the first processor will start its fourth iteration after the third iteration writes output to the FIFO buffer, delaying the execution of the fourth iteration slightly. For the sake of simplicity, such delays are ignored in Fig. 4.1b. After the first output from the pipelined MPSoC at 1,756 clock cycles, the third processor waits for processor 2's second iteration's output. Thus, after the first output from the pipelined MPSoC which corresponds to the filling of the pipeline, subsequent outputs are available every 878 clock cycles which is the critical processor's latency (processor 2). Following this line of reasoning, second and third outputs from the pipelined MPSoC will be available at $1,756 + 878 = 2,634$ and $2,634 + 878 = 3,512$ clock cycles respectively, as marked in Fig. 4.1b. From this observation, the execution time of the pipelined MPSoC will be $1,756 + (7 - 1) \times 878 = 7{,}024$ clock cycles. This concept can be generalised as follows to estimate the execution time of a pipelined MPSoC:

$$\mathrm{E} = \mathbb{I}(s_1) + \sum_{i=1}^{M} \mathbb{L}^1(s_i) + (\mathrm{I} - 1) \times \mathbb{L}(s_c) + \mathbb{F}(s_M)$$

where,

- $\mathbb{I}$: Returns the time spent in the initial non-kernel operations of a stage, that is, the time spent until the start of the kernel operation.

- $\mathbb{F}$: Returns the time spent in the final non-kernel operations of a stage, that is, the time spent after the end of the kernel operation.
- $\mathbb{L}^1$: Returns the latency of the first iteration of a stage.
- $\mathbb{L}$: Returns the latency of a stage that is averaged over all the iterations except the first one. Note that in a stage with more than one processor, the maximum latency from amongst all the processors in that stage is returned. The $\mathbb{I}$, $\mathbb{F}$ and $\mathbb{L}^1$ functions described above handle stages with multiple processors similar to the $\mathbb{L}$ function.
- s_i and s_c: The ith and the critical stage of the pipelined MPSoC respectively.
- I: The number of iterations of the pipelined MPSoC.
- M: The total number of stages in the pipelined MPSoC.

The above model considers the following factors which contribute to the execution time of a pipelined MPSoC:

- Initialisation time of the first stage—$\mathbb{I}(s_1)$;
- Time to fill the empty pipeline (time to enter steady state)—$\sum_{i=1}^{M} \mathbb{L}^1(s_i)$;
- Time spent by the critical stage in steady state—$(\mathrm{I} - 1) \times \mathbb{L}(s_c)$; and,
- Finalisation time of the last stage—$\mathbb{F}(s_M)$.

The reason for using $\mathbb{L}^1$ instead of $\mathbb{L}$ for the time to enter steady state is that there will be more cache misses in the first iteration compared to the second one due to cold cache start.

The throughput of a pipelined MPSoC depends on the latency of the critical processor. More formally,

$$\mathrm{T} = \frac{1}{\mathbb{L}(s_c)}$$

In the running example, $\mathrm{T} = 1/878$ data units/clock cycle, which is also marked in Fig. 4.1b.

The calculation of latency of a pipelined MPSoC is not as simple as the calculation of throughput. In the running example, the pipelined MPSoC enters into the steady state when the first processor starts its fourth iteration, as each processor's execution sequence repeats itself afterwards. For example, the first processor starts its iteration which is then followed by a stall period, similar to the third processor's execution sequence, though with different latencies and stalling periods. In the steady state, three factors contribute to the latency of a pipelined MPSoC. Firstly, the number of clock cycles spent in a processor's execution sequence, which includes execution of one iteration and the following stall period, becomes equal to the latency of the critical processor if that processor appears before the critical processor in the pipelined MPSoC. For example, iteration 4 of processor 1 and the corresponding stall period overlaps with the second iteration of processor 2, taking the same number of clock cycles as the critical processor's latency. Thus, given this first observation, it will take $878 + 878 = 1{,}756$ clock cycles for a data unit to appear at the output of processor 2. In general terms, it will take $i_c \times \mathbb{L}(s_c)$ clock cycles where i_c is the index of the critical stage, starting from 1. The second factor which contributes to the latency of a pipelined MPSoC depends on the number of FIFO buffers present in the pipelined

MPSoC on the critical path, that is, between the first and the critical processor. This is because processors appearing before the critical one can start their iterations earlier and hence will be processing data units further in the data stream compared to the critical processor. For example, the fourth iteration of processor 1 starts at the same time as the second iteration of processor 2, which means that the fourth data unit cannot be processed by processor 2 until the second and third data units are cleared. Since it is assumed that the FIFO buffers can accommodate the output of one iteration, the delay introduced due to the early start of the first processor is equal to the number of FIFO buffers present between the first and critical processors, multiplied by the critical latency. For example, there is one buffer between the first and second processors, meaning that there will be a delay of one extra iteration of the critical processor for a data unit to reach the critical processor, after being processed by the first processor. This is also illustrated in Fig. 4.1b, where the output of the fourth iteration of processor 1 waits for the third iteration of processor 2, adding 878 clock cycles to the latency of the pipelined MPSoC. Thus, using the first and second observations, it will take $1,756 + 878 = 2,634$ clock cycles for a data unit to appear at the output of processor 2. In general terms, according to the first and second observations, it will take $i_c \times \mathbb{L}(s_c) + (i_c - 1) \times \mathbb{L}(s_c) = (2 \times i_c - 1) \times \mathbb{L}(s_c)$ clock cycles, as there will be $i_c - 1$ FIFO buffers present between the critical processor and the first processor.[1] Once the latency of a data unit to appear at the output of the critical processor is estimated, the rest of the time is contributed by all the processors appearing after the critical processor, being the third factor. For example, the output of the fourth iteration of processor 2 is available at 3,826 clock cycles from the start time, and is then processed by the third processor to produce the fourth output at 4,390 clock cycles, where the latency of the fourth data unit will be $4{,}390 - 1{,}192 = 3 \times 878 + 564 = 3{,}198$ clock cycles. To summarize, the latency of a pipelined MPSoC can be estimated as follows:

$$\mathrm{L} = (2 \times i_c - 1) \times \mathbb{L}(s_c) + \sum_{i=i_c+1}^{M} \mathbb{L}(s_i)$$

The timing analysis presented here ignored the variations in performance that may occur due to the reading and writing of a FIFO buffer simultaneously, since such variations are small in practice. A more accurate analysis could have been conducted, but would have further complicated the execution time and latency models with little additional benefit. Note that these analytical models are applicable to pipelined MPSoCs that implement applications with backward edges given those edges fulfil the dependency distance condition. As such, the processors will not

[1] Note that in real-time, the input data to first processor will be available at the rate equal to throughput of the pipelined MPSoC. In such a scenario, number of FIFO buffers between the first processor and critical processor will not affect the latency of the pipelined MPSoC because the input data will not be available to the first processor in advance. Hence, the second factor will not contribute to the latency of a real-time pipelined MPSoC.

wait unnecessarily and pipelined execution will continue normally, and hence no additional factors need to be considered in the analytical models.

The latency and throughput of multimedia applications can be estimated by representing the applications as Synchronous Data Flow (SDF) graphs. Since SDFs allow generic backward edges, Maximum Cycle Mean (MCM), Max-Plus algebra and state-space exploration based techniques are used to compute latency and throughput [17]. These techniques are slow [18] and are not feasible when billions of design points of a pipelined MPSoC need to be evaluated. Unlike SDFs, in this chapter, the application model is restricted by the dependency distance condition for backward edges, yet allowing enough flexibility for modelling of real-world multimedia applications. Exploitation of the dependency distance condition results in analytical models of the pipelined MPSoC that are linear equations in latencies of processors, and thus avoids the computation of maximum cycles and state-space exploration, making them suitable for rapid exploration of large design spaces. The author does not know any work that reports exploration of billions of design points for a multimedia application using SDFs.

4.2 Estimation Methods

The execution time, latency and throughput of a pipelined MPSoC's design point can be estimated using the analytical models, if the latencies of those particular processor configurations are known. Hence, there is no need for full-system, cycle-accurate simulations of all the design points because the latencies of individual processor configurations can be captured. In this section, two methods are proposed to estimate the latencies of individual processor configurations with minimal number of simulations.

4.2.1 PS Method (Pipelined MPSoC Simulation)

In the simulation of a pipelined MPSoC with sufficiently sized FIFO buffers, the stalls of non-critical processors are hidden in the latency of the critical processor (see Fig. 4.1b). This observation leads to the fact that simulation of a pipelined MPSoC with one combination of processor configurations can be used to record the *net computation* and *net communication* latencies of individual processor configurations used in that particular simulation. Hence, in PS method, a pipelined MPSoC is simulated with the first available configuration of each processor. Then, the next available configuration of each processor is used. Figure 4.2 illustrates the PS method for the pipelined MPSoC of Fig. 4.1a where the three processors have 10, 20 and 15 configurations respectively. The PS method allows simulation of each processor configuration at least once and hence captures the net computation and communication latencies of all the processor configurations. For the running example, 20

Pipelined MPSoC 's Simulation	P1's Configuration	P2's Configuration	P3's Configuration
1	1	1	1
2	2	2	2
⋮	⋮	⋮	⋮
9	9	9	9
10	10	10	10
11	10	11	11
⋮	⋮	⋮	⋮
20	10	20	15

Fig. 4.2 An example of PS method where the three processors of pipelined MPSoC in Fig. 4.1a have 10, 20 and 15 configurations respectively

simulations will be required which is equal to the maximum number of processor configurations from amongst all the processors in the pipelined MPSoC. Note that a naive method will simulate all the possible combinations of processor configurations, that is, $10 \times 20 \times 15 = 3{,}000$ simulations. Once the latencies of processor configurations are available, the analytical models of the pipelined MPSoC are used to estimate the performance of any of its design point (any combination of processor configurations).

4.2.2 PSP Method (Pipelined MPSoC Simulation and Processor Analytical Model)

Although the PS method dramatically reduces the number of simulations, it will be slow when one of the processors in the pipelined MPSoC has hundreds of configurations which is the typical case (see Sect. 4.3). The author exploits the fact that a processor configuration is a combination of an ISA and a cache configuration. Hence, in the PSP method, a subset of processor configurations (instead of all the processor configurations) is simulated to gather architectural parameters of the ISAs and cache statistics (cache hit and miss counts) to build a processor analytical model, which is then used to estimate the latencies of the rest of the processor configurations. Since the ISAs are far less than the cache configurations, only a small subset of processor configurations need to be simulated to capture ISAs' architectural parameters. Note that the processor analytical model proposed here shares fundamental concepts with [8, 13–16]; however, those concepts have been extended to estimate latencies of processor configurations in a pipelined MPSoC.

The execution time, t_e, of a sub-kernel on a processor can be broken down into two parts: the time to fetch the instructions and data, t_f; and the net time to execute the fetched instructions, t_{ne}. The fetching time of instructions and data depends on the memory hierarchy of the processor. The time to execute the fetched instructions depends on the underlying micro-architecture, data dependencies and the total number of instructions in the program. A typical processor with classical 5-stage pipeline, in-order issue, separate L1 instruction and data caches, and a single memory for both instructions and data (local memory of each processor in the pipelined MPSoC) is assumed. A write-through cache policy is also assumed.

The following terminology is introduced to explain the processor analytical model:

- L_{IH}: Instruction cache hit latency.
- L_{IM}: Instruction memory read latency.
- L_{DMR}: Data memory read latency. Since instructions and data are in the same memory, $L_{IM} = L_{DMR}$.
- L_{DMW}: Data memory write latency.
- C_{IH}: Instruction cache hit count.
- C_{IM}: Instruction cache miss count.
- C_{DMR}: Data cache read miss count.
- C_{DMW}: Data cache write miss count.
- N_I: Total number of instructions.

The following analytical model estimates the latency of a sub-kernel on a processor:

$$\begin{aligned} t_e &= t_f + t_{ne} \\ &= L_{IH} \times C_{IH} + (1 + L_{IM}) \times C_{IM} \\ &\quad + (1 + L_{DMR}) \times C_{DMR} + (1 + L_{DMW}) \times C_{DMW} \\ &\quad + NCPI \times N_I \end{aligned}$$

The first four factors provide an estimate of the memory fetch time for both instructions and data. In a typical 5-stage pipeline of a processor, instructions are fetched in stage 1 (Instruction Fetch stage) while data fetches are processed in stage 4 (Memory stage), thereby overlapping instruction and data fetches. The two factors $(1 + L_{IM}) \times C_{IM}$ and $L_{IH} \times C_{IH}$ account for instruction fetches in case of both instruction cache hits and misses. The instruction cache miss latency, L_{IM}, is incremented by one to account for the clock cycle needed to check whether cache access was a hit or a miss. The data hits are ignored because they will be overlapped with the instruction hits or misses due to the pipeline in the processor. However, data misses may not be perfectly overlapped with instruction hits and misses, and hence are taken into consideration. To make the model more accurate, read and write data misses are included separately as $(1 + L_{DMR}) \times C_{DMR}$ and $(1 + L_{DMW}) \times C_{DMW}$.

Once the time for instructions and data fetches is estimated, the rest of the time is due to the execution of the fetched instructions. The last factor estimates the *net execution time* by multiplying the Net Clock cycles Per Instruction (NCPI) with the total number of instructions. Note that the NCPI is not the actual CPI of the processor; the last factor estimates the net time to execute the instructions once they have been fetched with their corresponding data. Hence, the NCPI captures various micro-architectural events such as the overlapping of data misses with instruction hits and misses, and stalls due to the data dependencies. The value of NCPI remains fairly constant across different cache configurations of a given ISA executing a given sub-kernel because the effect of cache configurations is taken into account by the cache statistics (instruction and data caches' hit and miss counts). The fluctuations in the value of NCPI across the same ISA but with different cache configurations are due to overlapped fetches of instructions and data, and stalls resulting from the data dependencies. To accurately model such micro-architectural events, one needs to perform cycle-accurate simulation or use data-flow analysis techniques. Such events are condensed into the NCPI parameter to keep the processor analytical model simple (though the model has illustrated reasonable absolute accuracy and fidelity, see Sect. 4.4).

To find the value of NCPI, the analytical model is rearranged as:

$$NCPI = \frac{t_e - t_f}{N_I}$$

The author proposes to run cycle-accurate simulations of a few processor configurations (explained later in this section) that contain the same ISA but different cache configurations to record both the actual latencies and cache statistics. Using the recorded values, the average NCPI value of that particular ISA is calculated. The average NCPI value is then used to estimate the latencies of the same ISA with the rest of the cache configurations using cache statistics of those configurations. Cache statistics are captured using trace-based cache simulators [10–12] which are much faster than full-system, cycle-accurate simulators.

The choice of cache configurations that need to be simulated for an ISA will affect the accuracy of the analytical model. In this chapter, identical instruction and data caches starting from the first cache configuration until the last one are simulated. Such a policy captures the effects of all the individual instruction and data cache configurations and is based on the analysis presented in [14] where the authors empirically show that an application's performance on baseline configurations is the most significant predictor of its performance on other configurations. Consider that a processor has one ISA, and its instruction and data cache sizes are changed from 1 to 32 KB, then there will be $1 \times 6 \times 6 = 36$ processor configurations. The author simulates the ISA with both 1 KB instruction and data caches, 2 KB instruction and data caches and so on until 32 KB instruction and data caches, resulting in simulations of only 6 processor configurations. The values recorded from these 6 simulations are used to compute the average NCPI for the ISA. The latencies of this ISA and the rest of the 30 cache configurations (rest of the 30 processor configurations) are then estimated by utilising the average NCPI value and the cache statistics of those 30 configurations (obtained from a trace-based cache simulator) in the analytical model.

In case of more than one ISA for the processor, a similar process is applied to other ISAs.

Unlike the PS method where all the 36 processor configurations are simulated, the PSP method simulates only 6 processor configurations, which further reduces the number of cycle-accurate simulations. Note that the PSP method will be less accurate compared to the PS method as it uses a processor analytical model to estimate the latencies of the processor configurations instead of relying on pure cycle-accurate simulations. Once the latencies of processor configurations are available, the analytical models of the pipelined MPSoC are used to estimate the performance of any of its design point (any combination of processor configurations).

4.3 Experimental Methodology

Five pipelined MPSoCs were created for the multimedia applications of Fig. 3.1 using a commercial design environment from Tensilica. The Xtensa LX2 [5] processor provides an Application Specific Instruction set Processor (ASIP) platform for creation of processor configurations, and comes with Xtensa RB-2007.1 toolset that includes a C/C++ compiler, an Instruction Set Simulator (ISS), Xtensa PRocessor Extension Synthesis (XPRES) and XTensa Modeling Protocol (XTMP).

XPRES analyses the C code and automatically generates application specific custom instructions, which may consist of a combination of fused operations, FLIX instructions [19], specialised operations [20] and vector operations. XPRES can also generate different sets of custom instructions, reflecting different ISAs. These custom instructions are output in the Tensilica Instruction Extension (TIE) language, and are compiled through the TIE compiler for seamless integration because the C/C++ compiler will automatically exploit the new instructions without the need for modification of the code.

XTMP is a multi-processor simulation environment which enables instantiation of multiple processors, connecting them via FIFO buffers to realise pipelined MPSoCs. The FIFO buffers provide blocking pop and push functions to read from and write to the buffer. A pop from an empty buffer and a push to a full buffer stalls the processor. During the simulation of a pipelined MPSoC, such stalls are recorded to calculate the *net computation* and *net communication* latencies of the processor configurations used in that particular simulation. XTMP uses ISS, Xtensa LX2's cycle-accurate simulator, to generate cycle-accurate performance measures of the pipelined MPSoC.

The trace-based cache simulator proposed in [10] was used. For a given trace, their simulator outputs cache statistics (hit and miss counts) for all the instruction and data cache configurations. Cache parameters include cache size, line size and associativity.

The pipelined MPSoCs were created by assigning each sub-kernel of a multimedia application to one or more processors as illustrated in Fig. 3.2. For example, the ME sub-kernel of H.264Enc in Fig. 3.1 is assigned to three processors. After the

Table 4.1 Processor configurations (ISAs × cache configurations)

Stage	JPEGEnc1	JPEGEnc2	JPEGDec	MP3Enc	H.264Enc
1	4×36	5×36	8×36	7×36	6×36
2	4×36	5×36	8×36 8×36 8×36	7×36 7×36	6×36 4×36 4×36
3	11×36	7×36 7×36 7×36	7×36	9×36	5×36
4	4×36	7×36	-	9×36	5×36
5	7×36	4×36	-	-	7×36
6	4×36	-	-	-	5×36
Design Space	4.2×10^{13}	2.35×10^{16}	1.73×10^{12}	1.92×10^{12}	1.42×10^{18}

allocation of sub-kernels to processors, differing sets of custom instructions (differing ISAs) are generated for the processors using XPRES. Each set of custom instructions (each ISA) is combined with differing cache configurations to create processor configurations where both instruction and data caches' sizes are changed from 1 to 32 KB. These cache configurations were chosen to generate reasonable a number of processor configurations and are not a limitation; cache line size and associativity could also have been changed to further increase the number of processor configurations. Table 4.1 reports the number of configurations for all the processors in the pipelined MPSoCs. Columns 2–6 report the names of the pipelined MPSoCs while rows 2–7 report the number of processor configurations in a particular stage. For example, the processor in stage 2 (greyed row) of JPEGEnc1 has $4 \times 36 = 144$ configurations, where 4 is the number of ISAs and 36 is the number of cache configurations. Since ME sub-kernel of H.264Enc is assigned to 3 processors, the entry for stage 2 of H.264Enc contains the number of processor configurations for the 3 processors, which are separated by the spaces. Note that processor configurations were not generated for the EC/W sub-kernel of the H.264Enc because the design space was already very large. JPEGDec has only three stages, and thus stages 4–6 contain no data. The last row shows the total number of possible combinations of processor configurations (the total number of design points) for each of the pipelined MPSoCs where the design spaces range from 10^{12} to 10^{18} design points.

All experiments were conducted on a quad core machine running at 2.15 GHz with 8 Gb RAM.

4.4 Results and Analyses

4.4.1 Processor's Analytical Model

Firstly, the evaluation of the processor analytical model is presented. Figure 4.3 illustrates the ratio of estimated latencies to actual latencies (plotted on y-axis) for all the configurations of the processors (plotted on x-axis) in the first stages of the pipelined

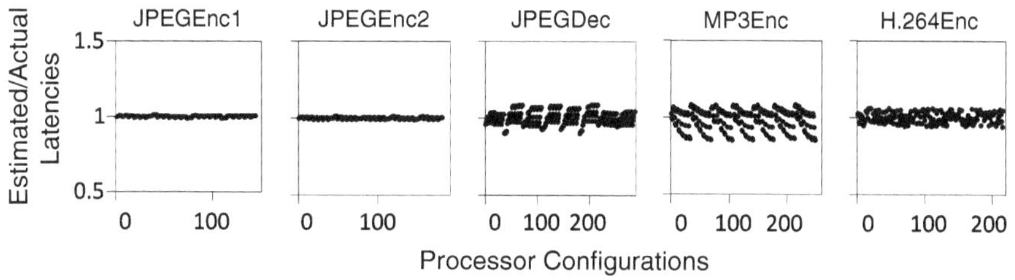

Fig. 4.3 Analysis of processor analytical model for the first stage's processor of each pipelined MPSoC

MPSoCs. If the ratio is 1, then the estimate is entirely accurate. The plots in Fig. 4.3 illustrate that the values are close to 1 which means that the latency estimates are reasonably accurate. Other processors in all the pipelined MPSoCs revealed similar findings.

A detailed analysis of the processor analytical model is reported in Table 4.2. The third and fourth columns report the absolute accuracy and fidelity[2] where both absolute error and fidelity are computed by comparing the actual latencies from cycle-accurate simulations of processor configurations to the estimated latencies from the processor analytical model. Note that P3.2 of JPEGEnc2 in Table 4.2 refers to the second processor in the third stage of the pipelined MPSoC. The empirical data in Table 4.2 reports the worst average and worst maximum absolute errors of 7.15 and 15.02 %, and minimum fidelity of 0.90 across all the processor configurations as highlighted in the table. These results show that the processor analytical model is reasonably accurate and hence is suitable for quick and early design space exploration of pipelined MPSoCs.

4.4.2 Pipelined MPSoC's Analytical Models and Estimation Methods

Table 4.3 reports detailed analysis of pipelined MPSoC's execution time, latency and throughput analytical models, and PS and PSP estimation methods. The second, third and fourth major columns report the results for execution time, latency and throughput analytical models. Minor columns in each of these major columns report absolute accuracy and fidelity of both PS and PSP methods. For the execution time analytical model, the PS method has worst average and worst maximum absolute errors of 3.83 and 6.89 % (MP3Enc) respectively across all the pipelined MPSoCs. In the PSP method, worst average and worst maximum absolute errors increased to

[2] Fidelity measures the correlation between the ordering of the actual and estimated values to quantify the similarity between the trends of actual and estimated values. A value close to 1 means that an analytical model has high fidelity. FM_ρ metric from [21] is used to compute fidelity due to its lower computational complexity.

Table 4.2 Detailed analysis of processor analytical model

Pipelined MPSoC	Processor	Absolute Error (%) Average	Absolute Error (%) Maximum	Fidelity
	P1.1	0.46	1.36	0.98
	P2.1	0.15	0.50	1.00
JPEGEnc1	P3.1	0.23	3.06	0.99
	P4.1	0.70	0.73	1.00
	P5.1	0.37	1.74	1.00
	P6.1	0.48	2.15	0.96
	P1.1	0.46	1.20	0.99
	P2.1	0.16	0.96	1.00
JPEGEnc2	P3.1	0.80	3.53	0.99
	P3.2	0.83	4.00	0.98
	P3.3	0.83	4.00	0.98
	P4.1	0.58	3.17	0.98
	P5.1	0.34	1.38	0.99
	P1.1	3.83	9.94	0.98
	P2.1	6.05	14.79	0.97
JPEGDec	P2.2	7.15	13.91	0.98
	P2.3	6.09	14.23	0.98
	P3.1	0.71	3.07	0.99
	P1.1	5.56	15.02	0.95
	P2.1	3.07	8.07	0.96
MP3Enc	P2.2	2.80	9.13	0.96
	P3.1	2.39	11.49	0.95
	P4.1	0.86	4.23	0.98
	P1.1	3.10	5.96	0.93
	P2.1	4.96	9.04	0.91
H.264Enc	P2.2	2.67	6.58	0.92
	P2.3	2.79	6.50	0.92
	P3.1	5.83	10.04	0.90
	P4.1	1.23	3.17	0.96
	P5.1	1.21	3.20	0.96
	P6.1	3.53	7.07	0.94

5.91 % (MP3Enc) and 13.21 % (JPEGDec) respectively. The minimum fidelity of the execution time analytical model dropped from 0.99 (JPEGEnc1) in the PS method to 0.93 (H.264Enc) in the PSP method. Similar results were found for both latency and throughput analytical models, which are reported in Table 4.3.

To summarise, among the three analytical models and all the pipelined MPSoCs, the worst average and worst maximum absolute errors of the PS method are 6.96 % (H.264Enc, throughput analytical model) and 12.95 % (H.264Enc, latency analytical model) with a minimum fidelity of 0.93 (H.264Enc, latency analytical model), as highlighted in Table 4.3. On the other hand, the PSP method has the worst average and

Table 4.3 Detailed analysis of pipelined MPSoC's execution time, latency and throughput analytical models, and PS and PSP estimation methods

Pipelined MPSoC	Execution Time						Latency						Throughput					
	Absolute Error (%)				Fidelity		Absolute Error (%)				Fidelity		Absolute Error (%)				Fidelity	
	Average		Maximum				Average		Maximum				Average		Maximum			
	PS	PSP	PS	PSP	PS	PSP	PS	PSP	PS	PSP	PS	PSP	PS	PSP	PS	PSP	PS	PSP
JPEGEnc1	2.28	5.00	5.91	9.11	0.99	0.96	0.88	1.87	3.78	8.59	0.98	0.94	1.71	6.28	5.87	10.75	0.99	0.94
JPEGEnc2	0.69	5.91	2.16	11.41	0.99	0.94	0.61	2.50	3.89	7.10	0.99	0.92	0.07	6.91	1.45	13.06	1.00	0.94
JPEGDec	0.21	5.08	1.29	13.21	0.99	0.98	1.73	6.31	5.33	15.66	0.98	0.88	0.28	6.76	1.90	18.67	0.99	0.98
MP3Enc	3.83	2.56	6.89	10.54	1.00	0.94	1.03	2.28	5.30	11.28	1.00	0.94	2.38	3.79	4.10	15.82	0.99	0.93
H.264Enc	1.47	3.11	2.72	8.33	0.99	0.93	6.28	6.83	12.95	17.56	0.93	0.93	6.96	7.56	7.41	13.36	0.99	0.93

worst maximum absolute errors of 7.56 % (H.264Enc, throughput analytical model) and 18.67 % (JPEGDec, throughput analytical model) with a minimum fidelity of 0.88 (JPEGDec, latency analytical model). The drop in accuracy and fidelity of the PSP method compared to the PS method is because it uses fewer cycle-accurate simulations and a processor analytical model instead of relying on pure cycle-accurate simulations as used in the PS method. Overall, the evaluation results indicate that execution time, latency and throughput analytical models, and the PS and PSP methods are reasonably accurate, and hence suitable for early design space exploration of pipelined MPSoCs.

4.4.3 Simulation Time of Estimation Methods

The advantage of the PSP method over the PS method is the reduction in simulation time due to the reduced number of cycle-accurate simulations, reported in Table 4.4. The second column reports the total number of design points. The time to simulate a design point depends on the pipelined MPSoC, and in our experiments simulation time of a design point varied from a few minutes to tens of minutes. Hence, simulation of the whole design space will take years and is not feasible.

The third and fourth major columns report the total number and time of simulations done in the PS and PSP methods. The PS method simulates each pipelined MPSoC for the maximum number of processor configurations from amongst all the processors in that pipelined MPSoC. Thus, 396 (11 × 36), 252 (7 × 36), 288 (8 × 36), 324 (9 × 36) and 252 (7 × 36) simulations were run for JPEGEnc1, JPEGEnc2, JPEGDec, MP3Enc and H.264Enc respectively.

Since there are billions of design points for each pipelined MPSoC, the absolute accuracy and fidelity of execution time, latency and throughput analytical models reported in Sect. 4.4.2 were computed using a few hundred design points. The author used the same design points that are simulated in the PS method because it ensures that all the individual processor configurations in a pipelined MPSoC are simulated at least once, and hence a reasonable evaluation can be conducted. More design points could have been used at the cost of increased simulation time.

Table 4.4 Simulation time of estimation methods

Pipelined MPSoC	Design Space	#Simulations		Simulation Time	
		PS	PSP	PS	PSP
JPEGEnc1	4.2×10^{13}	396	66	19 h	2 h
JPEGEnc2	2.35×10^{16}	252	42	15 h	1.5 h
JPEGDec	1.73×10^{12}	288	48	13 h	2 h
MP3Enc	1.68×10^{12}	252	42	2 days	16 h
H.264Enc	1.42×10^{18}	252	42	5 days	21 h

Table 4.5 Analysis of execution time analytical model proposed in [1]

Pipelined MPSoCs	Execution Time [1]					
	Absolute Error (%)				Fidelity	
	Average		Maximum			
	PS	PSP	PS	PSP	PS	PSP
JPEGEnc1	2.87	4.49	6.52	8.51	0.99	0.96
JPEGEnc2	1.10	5.59	2.59	10.97	0.99	0.94
JPEGDec	0.27	5.02	1.44	13.04	0.99	0.98
MP3Enc	19.89	15.55	23.19	22.75	0.99	0.94
H.264Enc	2.21	3.52	3.95	9.48	0.99	0.93

The PSP method simulates a subset of processor configurations to reduce simulation time. Recall from Sect. 4.2.2 that each ISA of a processor is simulated with identical instruction and data cache configurations to estimate the value of NCPI parameter. Since the cache sizes are changed from 1 to 32 KB, there are 6 identical instruction and data cache configurations; 1 KB instruction and data caches, 2 KB instruction and data caches and so on until 32 KB instruction and data caches. Hence, only 6 cycle-accurate simulations are run to estimate the value of NCPI of an ISA of a processor. The latencies of the rest of the combinations of cache configurations and the ISA are estimated using the processor analytical model. For example, the processor in the third stage of JPEGEnc1 has 11 ISAs and 36 cache configurations for each of those ISAs (from Table 4.1). Each ISA is simulated with 6 cache configurations, resulting in $11 \times 6 = 66$ simulations, which is maximum amongst all the other processors of JPEGEnc1. Hence, only 66 simulations are used by the PSP method for JPEGEnc1 compared to 396 simulations in the PS method. Compared to the PS method, the PSP method reduced simulation time (reported in the last major column of Table 4.4) from days to several hours because it reduced the number of simulations from hundreds to only tens.

4.4.4 Comparison to Prior Research

Shee et al. [1] also proposed an execution time analytical model for pipelined MPSoCs. Their model uses initialisation time of the first stage, time spent in critical stage and finalisation time of the last stage, ignoring the time to fill empty pipelined MPSoC (which can be computed from the latencies of first iteration, and is included in the author's execution time analytical model). In other words, their model focuses more on the steady state of a pipelined MPSoC. Table 4.5 reports the absolute accuracy and fidelity of their model when used with the PS and PSP estimation methods. Their execution time estimation is quite similar to the author's estimation (second major column of Table 4.3) except for MP3Enc. In the MP3Enc pipelined MPSoC, first iteration latencies have high magnitudes and the number of iterations is small which means that time to fill empty pipelined MPSoC is significant and cannot be ignored. That is why Shee's model exhibited high absolute errors for MP3Enc, which are highlighted in Table 4.5. In some cases (the PSP method for JPEGEnc1, JPEGEnc2 and MP3Enc), there are slight unexpected decreases in both average and maximum absolute errors of Shee's model compared to the author's model. This is because only a few hundred design points (design points that are simulated in the PS method) were used for calculation of absolute accuracy due to slow cycle-accurate simulations and huge design spaces. Use of only a few hundred design points also explains the decrease in both average and maximum absolute errors of the PSP method compared to PS method for MP3Enc in Table 4.5 in contrast to an obvious increase for the rest of the pipelined MPSoCs. Note that Shee's model exhibited the same fidelity as the author's model.

4.5 Summary

In this chapter, three analytical models and two estimation methods are proposed to aid quick design space exploration of pipelined MPSoCs. The pipelined MPSoC execution time, latency and throughput analytical models are linear in latencies of the individual processors. Hence, two estimation methods (PS and PSP) are proposed to quickly gather latencies of processor configurations with reduced number of slow, full-system, cycle-accurate simulations. The PS method simulates all the processor configurations once. On the other hand, the PSP method simulates a subset of processor configurations and then uses an analytical model of the processor to estimate latencies of processor configurations.

Experiments with five pipelined MPSoCs executing typical multimedia applications showed that the PS method had worst average and worst maximum absolute errors of 6.96 and 12.95 % with a minimum fidelity of 0.93. On the other hand, the PSP method had worst average and worst maximum absolute errors of 7.56 and 18.67 % with a minimum fidelity of 0.88. For design spaces ranging from 10^{12} to 10^{18} design points, the simulation time is reduced by several orders of magnitude; from days in the PS method to several hours in the PSP method. These results indicate that the proposed models and estimation methods are reasonably accurate, and hence suitable for rapid design space exploration of pipelined MPSoCs. The next chapter uses these analytical models to quickly explore a pipelined MPSoC's design space to optimise its area footprint under a performance constraint.

References

1. S.L. Shee, S. Parameswaran, Design methodology for pipelined heterogeneous multiprocessor system, in *DAC '07: Proceedings of the 44th annual conference on Design automation*, San Diego, pp. 811–816 2007

2. I. Karkowski, H. Corporaal, Design of heterogenous multi-processor embedded systems: applying functional pipelining, in *PACT '97: Proceedings of the 1997 International Conference on Parallel Architectures and Compilation Techniques*, IEEE Computer Society, San Francisco, 1997
3. H. Javaid, S. Parameswaran, Synthesis of heterogeneous pipelined multiprocessor systems using ilp: jpeg case study, in *CODES/ISSS '08: Proceedings of the 6th IEEE/ACM/IFIP international conference on Hardware/Software codesign and system synthesis*, ACM, New York, pp. 1–6 2008
4. H. Javaid, S. Parameswaran, A design flow for application specific heterogeneous pipelined multiprocessor systems, in *DAC '09: Proceedings of the 46th Annual Design Automation Conference*, ACM, New York, pp. 250–253 2009
5. Tensilica, Xtensa customizable processor. http://www.tensilica.com
6. M. Yourst, PTLsim: a cycle accurate full system x86–64 microarchitectural simulator, in *Performance Analysis of Systems and Software, ISPASS 2007. IEEE International Symposium on*, pp. 23–34 Aprl 2007
7. ARM, RealView ARMulator ISS. http://www.arm.com
8. R. Srinivasan, J. Cook, O. Lubeck, Performance modeling using monte carlo simulation, Comput. Archit. Lett. **5**, 38–41 (2006)
9. H. Javaid, A. Ignjatovic, S. Parameswaran, Rapid design space exploration of application specific heterogeneous pipelined multiprocessor systems. Trans. Comp.-Aided Des. Integ. Cir. Sys. **29**, 1777–1789 (2010)
10. M.S. Haque, J. Peddersen, A. Janapsatya, S. Parameswaran, Dew: a fast level 1 cache simulation approach for embedded processors with fifo replacement policy, in *DATE '10: Proceedings of the conference on Design, automation and test in*, Europe, 2010
11. J. Edler, M.D. Hill, Dinero iv trace-driven uniprocessor cache simulator. http://www.cs.wisc.edu/markhill/DineroIV/, 2004
12. N. Tojo, N. Togawa, M. Yanagisawa, T. Ohtsuki, Exact and fast l1 cache simulation for embedded systems, in *ASP-DAC '09: Proceedings of the 2009 Conference on Asia and South Pacific Design Automation*, IEEE Press, Piscataway, pp. 817–822 2009
13. L. Singleton, C. Poellabauer, K. Schwan, Monitoring of cache miss rates for accurate dynamic voltage and frequency scaling, in *Proceedings of the Multimedia Computing and Networking Conference (MMCN)*, 2005
14. B.C. Lee, D.M. Brooks, Accurate and efficient regression modeling for microarchitectural performance and power prediction, in ASPLOS-XII: Proceedings of the 12th International Conference on Architectural Support for Programming Languages and Operating Systems, ACM, New York, pp. 185–194 2006
15. P. Joseph, K. Vaswani, M. Thazhuthaveetil, Construction and use of linear regression models for processor performance analysis, in *Proceedings of the High-Performance Computer Architecture, The Twelfth International Symposium on*, pp. 99–108 2006
16. T.S. Karkhanis, J.E. Smith, A first-order superscalar processor model. SIGARCH Comput. Archit. News **32**(2), 338 (2004)
17. A. Ghamarian, M. Geilen, S. Stuijk, T. Basten, A. Moonen, M. Bekooij, B. Theelen, M. Mousavi, Throughput analysis of synchronous data flow graphs, in *Proceedings of the Application of Concurrency to System Design, ACSD 2006. Sixth International Conference on*, pp. 25–36 june 2006
18. A. Ghamarian, M. Geilen, T. Basten, S. Stuijk, Parametric throughput analysis of synchronous data flow graphs, in *Proceedings of the Design, Automation and Test in Europe, DATE '08*, pp. 116–121 March 2008
19. Tensilica, Flix: fast relief for performance-hungry embedded applications. http://www.tensilica.com/
20. Tensilica, XPRES generated specialized operations. http://www.tensilica.com/
21. H. Javaid, A. Ignjatovic, S. Parameswaran, Fidelity metrics for estimation models, in *Proceedings of the Computer-Aided Design (ICCAD), IEEE/ACM International Conference on*, pp. 1–8 Nov 2010

Chapter 5
Design Space Exploration of Pipelined MPSoCs

A pipelined MPSoC's stages need to be balanced for maximal utilisation of the processors to achieve high throughput with reduced area footprint and reduced power consumption. This chapter addresses the problem of optimising a pipelined MPSoC's area footprint. Like Chap. 4, each processor in the pipelined MPSoC has a number of configurations that trade-off performance with area footprint. Thus, a design point is one combination of processor configurations and the design space consists of all the combinations of processor configurations. The aim of this chapter is to quickly search the design space for the optimal design point (minimum area footprint) under either a performance constraint because such a constraint is often imposed on real-time multimedia applications. Two Integer Linear Programming (ILP) formulations for area footprint optimisation under an execution time constraint and a latency constraint are proposed in addition to an algorithm for area footprint optimisation under a throughput constraint.

Integer Linear Programming (ILP) is a widely used optimisation technique for heterogeneous MPSoCs, and has already been employed in several works [1–4] (see Sect. 2.3.1 for more references and details on ILP). However, those works focused on architectures other than the pipelined MPSoC. Jin et al. [5] addressed the problem of maximising the throughput of a multimedia application on a pipelined MPSoC with a fixed number of processors. Cong et al. [6] proposed exact algorithms to minimise latency and the number of processors in a pipelined MPSoC under a throughput constraint. Both these works [5, 6] did not consider processor customisation, and thus dealt with homogeneous pipelined MPSoCs only.

The works in [7–9] addressed the problem of processor customisation (selection of custom instructions or selection of processor configurations) in a pipelined MPSoC. Shee et al. [7] proposed a heuristic to maximise pipelined MPSoC's execution time improvement per area increase ratio compared to a single processor system. Thus, Shee et al. did not consider performance constraints that are typical of real-time multimedia applications. Two works inspired from the proposals of this chapter have been published recently [8, 9]. Bordoli et al. [9] considered variations in processor latencies during customisation of the processors. Their objective was to minimise

H. Javaid and S. Parameswaran, *Pipelined Multiprocessor System-on-Chip for Multimedia*, DOI: 10.1007/978-3-319-01113-4_5,

variation in throughput under an area footprint constraint. Chen et al. [8] explored simultaneous mapping and processor customisation with variable number of processors in the pipelined MPSoC. Their aim was to minimise MPSoC's area under a throughput constraint, but a latency constraint was not considered.

5.1 Problem Statement

A pipelined MPSoC where the application sub-kernels have already been mapped onto the processors is represented as a directed graph, PM:

$$PM = (P, F)$$

Each node in the set P is a processor, denoted as:

$$P = \{m.n \ : 1 \leq m \leq M, 1 \leq n \leq N_m\}$$

where M is the number of stages in the pipelined MPSoC and N_m is the number of processors in the mth stage. The processor $m.n$ is the nth processor in the mth stage of the pipelined MPSoC. Each edge in the set F is a FIFO buffer, denoted as:

$$F = \{(m.n : i.j) : 1 \leq m, i \leq M, 1 \leq n \leq N_m, 1 \leq j \leq N_i\}$$

For example, the FIFO buffer between processors 2.1 and 3.1 in a pipelined MPSoC will be denoted as 2.1:3.1. The execution time, latency and throughout of a pipelined MPSoC are denoted as E, L and T, and are calculated using the analytical models proposed in Sect. 4.1. The area of the pipelined MPSoC is the summation of the area of all the processors and FIFOs, calculated as:

$$A = \sum_{m=1}^{M} \sum_{n=1}^{N_m} \left[\mathbb{A}(m.n) + \sum_{i=1}^{M} \sum_{j=1}^{N_i} \mathbb{A}(m.n : i.j) \right]$$

where the function $\mathbb{A}$ returns the area of the processors and FIFO buffers.

The processors in the pipelined MPSoC have a number of configurations, trading-off their latency with area footprint. The configurations of a processor $m.n$ are denoted as:

$$C = \{m.n_o : 1 \leq m \leq M, 1 \leq n \leq N_m, 1 \leq o \leq O_{m,n}\}$$

where $O_{m,n}$ is the total number of configurations available for the processor $m.n$. For example, the second configuration of processor 2.1 is denoted as 2.1_2. Each processor configuration $m.n_o$ is annotated with a 5-tuple number denoting the initial time, latency of first iteration, average latency and final time of executing the assigned sub-kernel(s) on that particular configuration, and the area of that particular configuration. The latency of a processor configuration includes both the *net computation* and *net communication* latencies as explained in Sect. 4.2.1. The functions $\mathbb{I}$, $\mathbb{L}^1$, $\mathbb{L}$, $\mathbb{F}$ and $\mathbb{A}$ return the aforementioned tuples of a processor configuration.

Given the above definitions, the optimisation problem can be stated as: *For a pipelined MPSoC where each processor has a number of configurations, the goal is to select one configuration for each processor so that the area footprint of the pipelined MPSoC is minimum and its execution time (or latency or throughput) satisfies the execution time (or latency or throughput) constraint E_c (or L_c or T_c), provided by the designer.* Typically, execution time and latency constraints are provided as an upper bound, while the throughput constraint is specified as a lower bound. Here, the author assumes the throughput constraint to be an upper bound as well, that is, a constraint on the latency of the critical processor in the pipelined MPSoC (which can be calculated by inverting the throughput constraint provided by the designer). Thus, all the constraints (E_c, L_c and T_c) are assumed to be in clock cycles. The following sections describe the techniques proposed to optimise the area footprint under these constraints.

5.2 Optimisation Under an Execution Time Constraint

The processor configuration selection problem under an execution time constraint is formulated as a binary ILP problem in the following way:

5.2.1 Variables

Binary variables are used to determine the selection of processor configurations:

- $x_{m,n,o}$ variables are used to select one configuration per processor. A variable $x_{m,n,o}$ equals 1 if the configuration o of processor $m.n$ is selected, otherwise equals 0.
- $s_{m,n,o}$ variables are used to select one configuration per pipeline stage. A stage with more than one processor will have one configuration selected for each of the processors; however, only one of those (the one with maximum latency) configurations can be selected as the stage configuration for calculation of the pipelined MPSoC's execution time. A variable $s_{m,n,o}$ equals 1 if the configuration o of processor $m.n$ is also selected as the configuration of stage m, otherwise equals 0. These variables are only used for stages that contain more than one processor because configuration of a stage with only one processor will be the configuration selected for the only processor in that stage.

5.2.2 Objective Function

The objective function of the optimisation problem is to minimise the pipelined MPSoC's area footprint, which can be written as:

$$\text{Minimise} \sum_{m=1}^{M} \sum_{n=1}^{N_m} \sum_{o=1}^{O_{m,n}} \mathbb{A}(m.n_o) x_{m,n,o}$$

Note that the area of the FIFO buffers is ignored here because their area is constant for a given pipelined MPSoC and hence does not affect the optimisation problem.

5.2.3 Constraints

Various constraints applicable to the processor configuration selection problem are listed below:

1. Only one configuration can be selected for a processor:

$$\sum_{o=1}^{O_{m,n}} x_{m,n,o} = 1 \quad \forall\, m, n$$

2. For stages with more than one processor, only one processor configuration can be selected as the stage configuration:

$$\sum_{n=1}^{N_m} \sum_{o=1}^{O_{m,n}} s_{m,n,o} = 1 \quad \forall\, m \text{ where } N_m > 1$$

3. The configuration of a stage with more than one processor must be one of the processor configurations selected using $x_{m,n,o}$ variables:

$$s_{m,n,o} - x_{m,n,o} \leq 0 \quad \forall\, m, n, o \text{ where } N_m > 1$$

4. Of the selected processor configurations in a stage with more than one processor, the configuration with maximum latency must be selected as the stage configuration. The following constraint compares the latencies of the nth and jth processors in the mth stage:

$$\max_{1 \leq o \leq O_{m,n}} \{\mathbb{L}(m.n_o)\} \times \left[1 - \sum_{o=1}^{O_{m,j}} s_{m,j,o} \right] + \sum_{m=1}^{O_{m,j}} \mathbb{L}(m.j_o) s_{m,j,o} \geq \sum_{o=1}^{O_{m,n}} \mathbb{L}(m.n_o) x_{m,n,o}$$

$$\forall\, n, j \text{ where } 1 \leq n, j \leq N_m, j \neq n \text{ and } N_m > 1$$

There will be $N_m - 1$ such constraints for each processor in stage m as it has to be compared with every other processor in that stage. Thus, in total there will be $N_m(N_m - 1)$ such constraints for stage m to ensure that the stage configuration is the one with maximum latency.

5. The execution time of the pipelined MPSoC (calculated using analytical model proposed in Sect. 4.1) must be less than or equal to the execution time constraint E_c. Since any processor can be critical in the pipelined MPSoC, the following constraint is used considering processor $m_c.n_c$ critical:

$$\sum_{n=1}^{N_1} \sum_{o=1}^{O_{1,n}} \mathbb{I}(1.n_o) s_{1,n,o} + \sum_{m=1}^{M} \sum_{n=1}^{N_m} \sum_{o=1}^{O_{m,n}} \mathbb{L}^1(m.n_o) s_{m,n,o}$$
$$+ (\mathrm{I} - 1) \times \sum_{o=1}^{O_{m_c,n_c}} \mathbb{L}(m_c.n_{co}) x_{m_c,n_c,o} + \sum_{n=1}^{N_M} \sum_{o=1}^{O_{M,n}} \mathbb{F}(M.n_o) s_{M,n,o} \le E_c$$
$$\forall\, m_c, n_c \text{ where } 1 \le m_c \le M,\ 1 \le n_c \le N_{m_c}$$

For stages with only one processor (that is, $N_m = 1$), $s_{m,n,o}$ variables are replaced by $x_{m,n,o}$ in the above constraint because $s_{m,n,o}$ variables are only used for stages with more than one processor (that is, $N_m > 1$). The above constraint is repeated considering each processor critical at a time, resulting in $\sum_{m=1}^{M} \sum_{n=1}^{N_m}$ many constraints where the selection of the critical processor is embedded into these constraints.

The output of the above binary ILP is a set of processor configurations with one configuration per processor, which will be optimal due to the use of binary ILP. For area footprint optimisation under an execution time constraint, the work in [10] reports more advanced techniques to prune the design space and relax the ILP formulation or use a heuristic to quickly search for a near-optimal solution.

5.3 Optimisation Under a Latency Constraint

The processor configuration selection problem under a latency constraint is formulated as a binary ILP problem in the following way:

5.3.1 Variables

Binary variables are used to determine the selection of processor configurations:

- $x_{m,n,o}$ variables are used to select one configuration per processor. A variable $x_{m,n,o}$ equals 1 if the configuration o of processor $m.n$ is selected, otherwise equals 0.

- $s_{m,n,o}$ variables are used to select one configuration per pipeline stage. A stage with more than one processor will have one configuration selected for each of the processors; however, only one of those (the one with maximum latency) configurations can be selected as the stage configuration for calculation of the pipelined MPSoC's latency. A variable $s_{m,n,o}$ equals 1 if the configuration o of processor $m.n$ is also selected as the configuration of stage m, otherwise equals 0. These variables are only used for stages that contain more than one processor because configuration of a stage with only one processor will be the configuration selected for the only processor in that stage.

5.3.2 Objective Function

The objective function of the optimisation problem is to minimise the pipelined MPSoC's area footprint, which can be written as:

$$\text{Minimise} \sum_{m=1}^{M} \sum_{n=1}^{N_m} \sum_{o=1}^{O_{m,n}} \mathbb{A}(m.n_o) x_{m,n,o}$$

Note that the area of the FIFO buffers is ignored here because their area is constant for a given pipelined MPSoC and hence does not affect the optimisation problem.

5.3.3 Constraints

Various constraints applicable to the processor configuration selection problem are listed below:

1. Only one configuration can be selected for a processor:

$$\sum_{o=1}^{O_{m,n}} x_{m,n,o} = 1 \qquad \forall\, m,n$$

2. For stages with more than one processor, only one processor configuration can be selected as the stage configuration:

$$\sum_{n=1}^{N_m} \sum_{o=1}^{O_{m,n}} s_{m,n,o} = 1 \quad \forall\, m \text{ where } N_m > 1$$

3. The configuration of a stage with more than one processor must be one of the processor configurations selected using $x_{m,n,o}$ variables:

$$s_{m,n,o} - x_{m,n,o} \leq 0 \quad \forall m, n, o \text{ where } N_m > 1$$

4. Of the selected processor configurations in a stage with more than one processor, the configuration with maximum latency must be selected as the stage configuration. The following constraint compares the latencies of the nth and jth processors in the mth stage:

$$\max_{1 \leq o \leq O_{m,n}} \{\mathbb{L}(m.n_o)\} \times \left[1 - \sum_{o=1}^{O_{m,j}} s_{m,j,o}\right] + \sum_{m=1}^{O_{m,j}} \mathbb{L}(m.j_o) s_{m,j,o} \geq \sum_{o=1}^{O_{m,n}} \mathbb{L}(m.n_o) x_{m,n,o}$$

$$\forall n, j \text{ where } 1 \leq n, j \leq N_m, j \neq n \text{ and } N_m > 1$$

There will be $N_m - 1$ such constraints for each processor in stage m as it has to be compared with every other processor in that stage. Thus, in total there will be $N_m(N_m - 1)$ such constraints for stage m to ensure that the stage configuration is the one with maximum latency.
5. The latency of the pipelined MPSoC (calculated using analytical model proposed in Sect. 4.1) must be less than or equal to the latency constraint L_c. Since any processor can be critical in the pipelined MPSoC, the direct use of the pipelined MPSoC's latency analytical model results in a non-linear constraint due to the product factor. In this monograph, such a constraint is lineraised by considering one of the processors critical at a time, leading to the following constraint:

$$(2 \times m_c - 1) \times \sum_{n=1}^{N_{mc}} \sum_{o=1}^{O_{mc,n}} \mathbb{L}(m_c.n_o) s_{m_c,n,o} + \sum_{m=m_c+1}^{M} \sum_{n=1}^{N_m} \sum_{o=1}^{O_{m,n}} \mathbb{L}(m.n_o) s_{m,n,o} \leq L_c$$

where m_c refers to the stage of the processor currently being considered critical. For stages with only one processor (that is, $N_m = 1$), $s_{m,n,o}$ variables are replaced by $x_{m,n,o}$ in the above constraint because $s_{m,n,o}$ variables are only used for stages with more than one processor (that is, $N_m > 1$). The following constraint is also added to make sure that the configurations selected for non-critical processors have lower latencies than the critical processor's latency, where the critical processor is referred to as $m_c.n_c$:

$$\sum_{o=1}^{O_{m,n}} \mathbb{L}(m.n_o) x_{m,n,o} \leq \sum_{o=1}^{O_{mc.nc}} \mathbb{L}(m_c.n_{co}) x_{m_c,n_c,o} \quad \forall m, n \text{ and } (m, n) \neq (m_c, n_c)$$

Since the binary ILP formulation described above considers only one of the processors to be critical, one instance of such a formulation provides solution for that particular critical processor only, and thus not the whole optimisation problem. Therefore, $\sum_{m=1}^{M} N_m$ instances of the binary ILP formulation are run, where these instances successively consider each processor as the critical processor in the pipelined MPSoC. Then, the solution with minimum area footprint from amongst the solutions of all the binary ILP instances is selected. Algorithmically, the optimisation approach is

Algorithm 1: Optimisation Under a Latency Constraint

```
1 OptimalSol = NULL;
  // One by one consider each processor as critical in the
  pipelined MPSoC
2 for m=1 to M do
3     for n=1 to N_m do
4         CurrentSol = SolveILP(m, n, L_c);
5         if CurrentSol's area footprint < OptimalSol's area footprint then
6             OptimalSol = CurrentSol;
7 return OptimalSol;
```

shown in Algorithm 1. The two for-loops starting at lines 2 and 3 traverse all the processors in the pipelined MPSoC, calling the function $SolveILP(m, n, L_c)$ with processor $m.n$ to be considered as the critical processor ($m_c = m$ and $n_c = n$ in constraint 5 of the binary ILP formulation). The output of the algorithm is a set of processor configurations with one configuration per processor. Since binary ILP is used, the selected design point will be optimal.

Although the binary ILP formulation needs to be run multiple times, the number of processors in a pipelined MPSoC is typically in the order of tens. Thus, the running time of Algorithm 1 will be reasonable (refer to Sect. 5.7). Alternatively, the optimisation problem could have been formulated as a non-linear problem, which is beyond the scope of this monograph.

5.4 Optimisation Under a Throughput Constraint

Optimisation of a pipelined MPSoC's area footprint under a throughput constraint is not as complex as its latency constrained optimisation. The algorithm is shown in Algorithm 2. Intuitively, a throughput constraint T_c on a pipelined MPSoC means that none of the processors in the pipelined MPSoC can have latency greater than T_c. Thus, the algorithm traverses all the configurations of all the processors (lines 1–5), and deletes any configuration with latency greater than T_c. After this pruning phase, the algorithm selects the configuration with the minimum area footprint for a processor from its remaining configurations, repeating the process for all the processors in the pipelined MPSoC. Such a selection results in minimum area footprint of the pipelined MPSoC because its area footprint is a linear summation of the area of individual processors. Thus, Algorithm 2 outputs one configuration per processor where the selected design point is optimal. The algorithm traverses all the processor configurations only once, resulting in a complexity of $O(\sum_{m=1}^{M} \sum_{n=1}^{N_m} o_{m,n})$.

Algorithm 2: Optimisation Under a Throughput Constraint

// Prune processor configurations with latency greater than T_c
for m=*1 to* M **do**
  for n=*1 to* N_m **do**
    for o=*1 to* $O_{m,n}$ **do**
      if $\mathbb{L}(m.n_o) > T_c$ **then**
        Delete processor configuration $m.n_o$

// select minimum area footprint configurations
for m=*1 to* M **do**
  for n=*1 to* N_m **do**
    Select configuration with minimum area for processor $m.n$ from the remaining configurations

5.5 Discussion

The pipelined MPSoC used in this monograph does not restrict processors to run at the same frequencies. If the processors are running at different frequencies, then their latencies in clock cycles cannot be just added to compute the execution time and latency of the pipelined MPSoC. In such a scenario, the latency of a processor in clock cycles should be converted to actual time, by dividing it by the frequency of that processor. Furthermore, the execution time and latency constraints should be provided as actual times rather than clock cycles. However, these steps will not change the binary ILP formulation and Algorithm 1. Likewise, for area footprint optimisation under a throughput constraint, both the processor latencies and throughput constraint should be converted to actual time. This will again not change Algorithm 2. Hence, these simple modifications can extend the proposed optimisation methods to pipelined MPSoCs with processors running at different frequencies.

5.6 Experimental Methodology

Design spaces of five pipelined MPSoCs, created in Chap. 4, were explored using the proposed optimisation methodologies. The number of configurations for each processor and the total design points are reported in Table 4.1 where the design spaces ranged from 10^{12} to 10^{18} design points. The latencies of the processor configurations were gathered using the PS method described in Sect. 4.2.1. The area of each processor configuration was measured in the number of gates and included the area of the base processor, custom instructions, and instruction and data caches.

A commercial linear programming solver, CPLEX [11], was used to solve the binary ILP formulation. CPLEX reads an input file in LP format and outputs the values of variables in a text file. The proposed algorithms were programmed in

Perl and integrated with Tensilica's design environment to automate the exploration process. All the experiments were conducted on a 2.15 GHz quad core machine with 8GB RAM.

5.7 Results and Analyses

5.7.1 Pareto Fronts

Figures 5.1, 5.2, 5.3, 5.4 and 5.5 show the results of design space exploration of the pipelined MPSoCs for execution time, latency and throughput. For the sake of simplicity, the values on the axes are omitted.

Figures 5.1a, 5.2a, 5.3a, 5.4a and 5.5a show the Pareto front of each pipelined MPSoC, where the execution time is plotted on the y-axis while the area is plotted on the x-axis. These Pareto fronts were obtained by specifying different execution time constraints, spanning the whole design space, and obtaining the optimal design

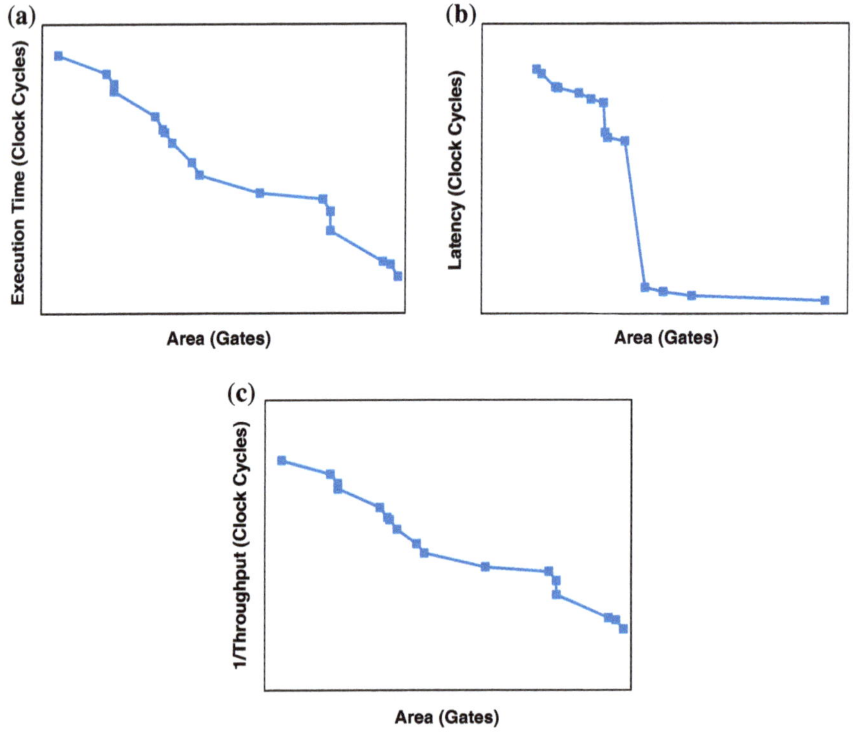

Fig. 5.1 Pareto fronts of JPEGEnc1: **a** execution time, **b** latency, and **c** throughput against area footprint

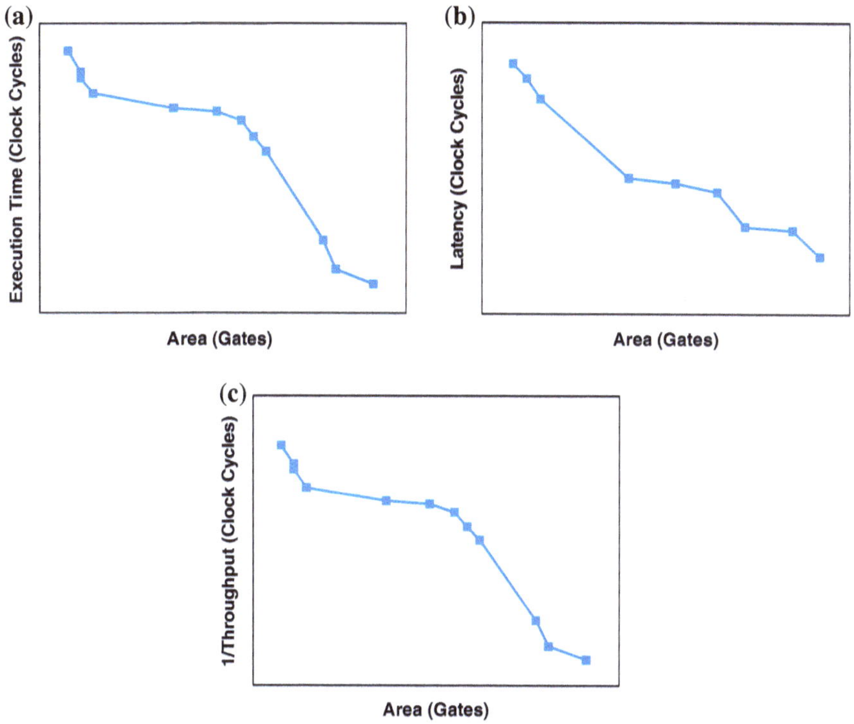

Fig. 5.2 Pareto fronts of JPEGEnc2: **a** execution time, **b** latency and **c** throughput against area footprint

point for each of the individual constraints. For example, the JPEGEnc1 design space was explored by providing execution time constraints from 3.5 million to 5.5 million clock cycles in steps of 100,000 clock cycles. The design spaces of other pipelined MPSoCs were explored similarly, though with different ranges and steps. Figures 5.1b, 5.2b, 5.3b, 5.4b and 5.5b show the Pareto fronts for all the pipelined MPSoCs with respect to latency while Figs. 5.1c, 5.2c, 5.3c, 5.4c and 5.5c show the Pareto fronts for *1/Throughput*, which translates to the latency of the critical processor in the pipelined MPSoC. A decrease in $1/Throughput$ of a pipelined MPSoC means a tighter bound on the critical latency, which will require more resources, and thus will increase the area footprint of a pipelined MPSoC as depicted in the Pareto fronts.

5.7.2 Exploration Time

An important concern while exploring the design space is the time taken to obtain the Pareto front. Table 5.1 reports the time to find the Pareto front of each pipelined MPSoC for execution time, latency and throughput. The second, third and fourth

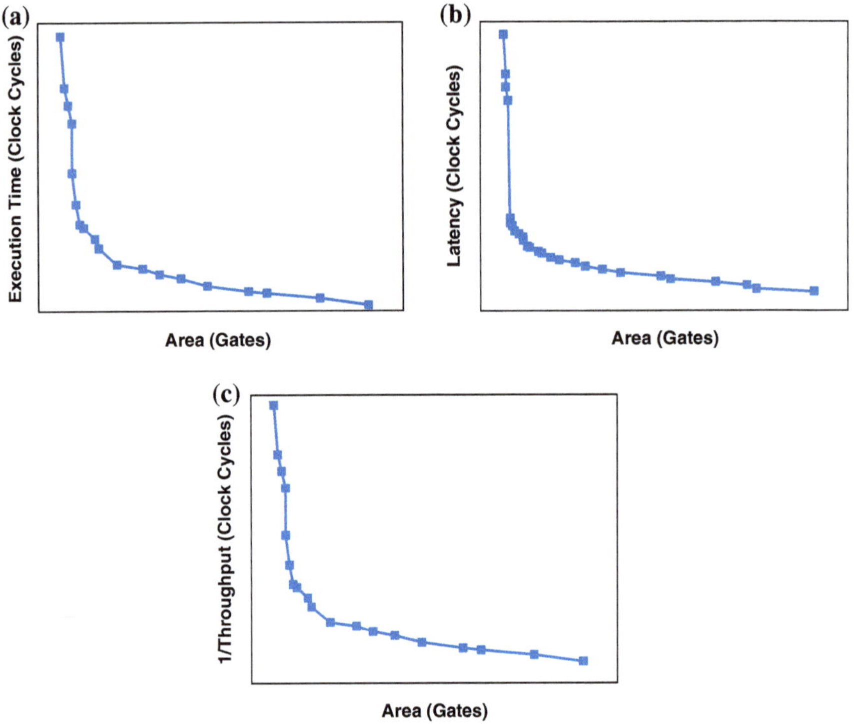

Fig. 5.3 Pareto fronts of JPEGDec: **a** execution time, **b** latency and **c** throughput against area footprint

Table 5.1 Exploration time to obtain Pareto fronts

Pipelined MPSoC	Execution time (s)	Latency (s)	Throughput (s)
JPEGEnc1	51	39	2
JPEGEnc2	77	81	3
JPEGDec	248	218	3
MP3Enc	264	226	2
H.264Enc	353	409	14

columns, titled execution time, latency and throughput, refer to area footprint optimisation under an execution time, a latency and a throughput constraint respectively. Since optimisation under an execution time constraint and a latency constraint uses binary ILP to find the optimal design point, their exploration time will be higher than that of Algorithm 2. The maximum time to find the Pareto front with respect to execution time and latency was less than 6 min (MP3Enc) and seven minutes (H.264Enc) respectively. With respect to throughput, the maximum time to find the Pareto front was 14 s, occurring for H.264Enc again.

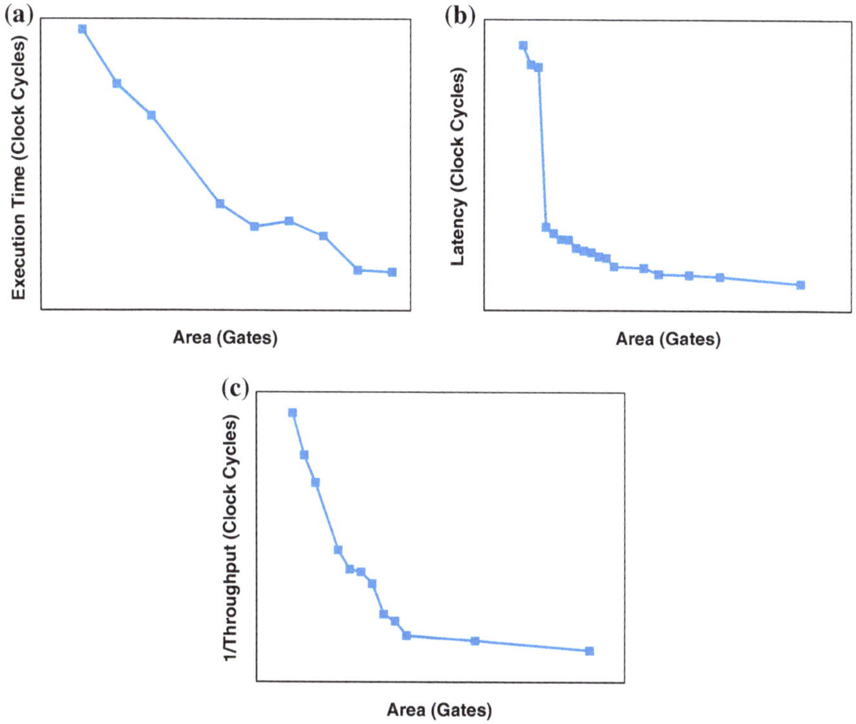

Fig. 5.4 Pareto fronts of MP3Enc: **a** execution time, **b** latency, and **c** throughput against area footprint

Note that the reported exploration times depend on the number of constraints used while exploring the design space, in addition to the complexity of binary ILP when an execution time constraint is used or Algorithm 1 (which depends on the complexity of binary ILP) when a latency constraint is used or the complexity of Algorithm 2 when a throughput constraint is used. In these experiments, a minimum of 50 and a maximum of 200 execution time constraints were used for design space exploration of the pipelined MPSoCs. Likewise, a minimum of 36 and a maximum of 82 latency constraints were used. For throughput constrained design space exploration, at least 44 throughput constraints were used for each pipelined MPSoC, with a maximum of 240 for H.264Enc. This shows that the proposed optimisation methods can handle exploration of reasonably large design spaces (with 10^{12} to 10^{18} design points), finding the Pareto front in a few minutes for three different performance metrics. Once these Pareto fronts are available, designers can trade-off the execution time, latency or the throughput with the area footprint by choosing an appropriate set of processor configurations for a pipelined MPSoC.

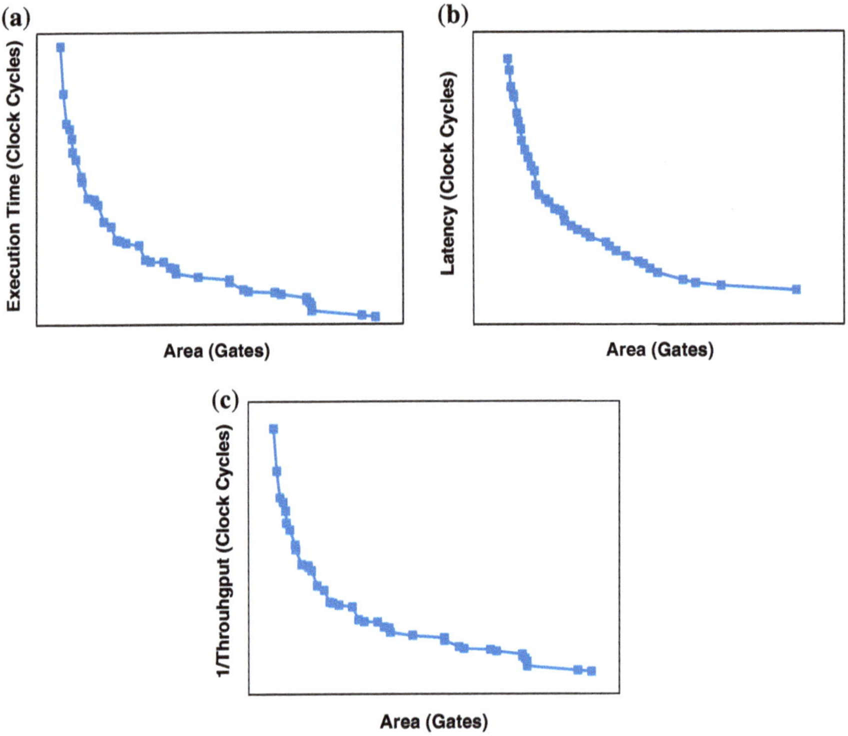

Fig. 5.5 Pareto fronts of H.264Enc: **a** execution time, **b** latency and **c** throughput against area footprint

5.7.3 JPEG Encoder Case Study

The JPEG encoder application in Fig. 3.1 was partitioned in two differing ways to compare alternative implementations of it on pipelined MPSoCs. Table 5.2 shows the comparison of JPEGEnc1 with JPEGEnc2. The second, third and fourth major columns refer to the optimal design point obtained under an execution time, a latency and a throughput constraint respectively where the constraints are in clock cycles. The term A stands for the area footprint of the selected design point, measured in number of gates. Comparing JPEGEnc1 with JPEGEnc2, for the same execution time constraint, the area of JPEGEnc2 (662,334 gates) is smaller than the area of JPEGEnc1 (672,288 gates), resulting in a 1.48 % reduction. Likewise, for the same latency and throughput constraints, JPEGEnc2 had an area reduction of 5.68 and 2.44 % respectively. This is because the three processors in the third stage process Y, Cb and Cr components of a macroblock in parallel, and thus increase the performance of the pipelined MPSoC. Therefore, simpler processor configurations in JPEGEnc2 can be used to achieve the same performance as of JPEGEnc1, resulting in lower area footprint of JPEGEnc2. Since the exploration time of the proposed

Table 5.2 Comparison of JPEGEnc1 and JPEGEnc2

Pipelined	Execution time constrained		Latency constrained		Throughput Constrained	
MPSoC	E_c	A	L_c	A	T_c	A
JPEGEnc1	5,300,000	672,288	36,000	700,104	5,800	678,953
JPEGEnc2	5,300,000	662,334	36,000	660,286	5,800	662,334

optimisation methods is in minutes, a designer can quickly compare and evaluate different pipelined MPSoCs for the same application. However, application partitioning and mapping need to be done either manually or semi-automatically using one of the several techniques discussed in Chaps. 2 and 3.

5.8 Summary

In this chapter, the author proposed several techniques to quickly search an optimal design point (minimum area footprint) of a pipelined MPSoC where its design space consisted of differing combinations of processor configurations. Since performance requirements of multimedia applications put constraints on the design of pipelined MPSoCs, area footprint optimisation was done under these constraints. For five pipelined MPSoCs, the exploration time to find Pareto fronts of each of those pipelined MPSoCs was less than seven minutes when their design spaces contained at least 10^{12} design points. This illustrates the applicability of the proposed methods to quickly optimise area footprint of performance constrained pipelined MPSoCs.

References

1. B.K. Dwivedi, A. Kumar, M. Balakrishnan, Synthesis of application specific multiprocessor architectures for process networks, in *VLSID '04: Proceedings of the 17th International Conference on VLSI Design* (Washington, DC, USA), p. 780, IEEE Computer Society, 2004
2. S.-R. Kuang, C.-Y. Chen, R.-Z. Liao, Partitioning and pipelined scheduling of embedded system using integer linear programming, in *ICPADS '05: Proceedings of the 11th International Conference on Parallel and Distributed Systems Workshops (ICPADS'05)* (Washington, DC, USA), pp. 37–41, IEEE Computer Society, 2005
3. J. Wu, J. Williams, N. Bergmann, An ilp formulation for architectural synthesis and application mapping on fpga-based hybrid multi-processor soc, in *International Conference on Field Programmable Logic and Applications, FPL 2008*, pp. 451–454, 2008
4. C.-L. Sotiropoulou, S. Nikolaidis, Ilp formulation for hybrid fpga mpsocs optimizing performance, area and memory usage, in *18th IEEE International Conference on Electronics, Circuits and Systems (ICECS)*, pp. 748–751, 2011
5. Y. Jin, N. Satish, K. Ravindran, K. Keutzer, An automated exploration framework for fpga-based soft multiprocessor systems, in *CODES+ISSS '05: Proceedings of the 3rd IEEE/ACM/IFIP International Conference on Hardware/Software Codesign and System Synthesis* (ACM, New York), pp. 273–278, 2005

6. J. Cong, G. Han, W. Jiang, Synthesis of an application-specific soft multiprocessor system, in *FPGA '07: Proceedings of the 2007 ACM/SIGDA 15th International Symposium on Field Programmable Gate Arrays* (ACM, New York), pp. 99–107, 2007
7. S.L. Shee, S. Parameswaran, Design methodology for pipelined heterogeneous multiprocessor system, in *DAC '07: Proceedings of the 44th Annual Conference on Design Automation*, pp. 811–816, 2007
8. L. Chen, N. Boichat, T. Mitra, Customized mpsoc synthesis for task sequence, in *Proceedings of the 2011 IEEE 9th Symposium on Application Specific Processors, SASP '11* (Washington, DC, USA), pp. 16–21, IEEE Computer Society, 2011
9. U. Bordoloi, H.P. Huynh, T. Mitra, S. Chakraborty, Design space exploration of instruction set customizable mpsocs for multimedia applications, in International Conference on Embedded Computer Systems (SAMOS), pp. 170–177, 2010
10. H. Javaid, A. Ignjatovic, S. Parameswaran, Rapid design space exploration of application specific heterogeneous pipelined multiprocessor systems. Trans. Comp.-Aided Des. Integ. Cir. Sys. **29**, 1777–1789 (2010)
11. IBM, "ILOG CPLEX Optimizer". Available at: http://www-01.ibm.com/

Chapter 6
Adaptive Pipelined MPSoCs

Multimedia applications exhibit variations in computational workload of their sub-kernels due to the adaptive nature of algorithms and input data. For example, the workload of the motion estimation sub-kernel in H.264 video encoder varies depending on the amount of motion in the incoming video frames. Therefore, to guarantee throughput at all times, pipelined MPSoCs have to be designed with worst-case parameters. For example, the optimisation methods proposed in Chap. 5 for design-time balancing of pipelined MPSoCs will use worst-case latencies of the processor configurations (that is, processor latencies are gathered by providing worst-case representative input data to the pipelined MPSoCs). Since worst-case pipelined MPSoCs lack adaptability to run-time variations in their computational workload, they suffer from inefficient resource utilisation and may result in high energy consumption under a dynamic workload. Let us examine the limitations of a worst-case pipelined MPSoC through a case study of the motion estimation in the H.264 video encoder.

6.1 Motivational Example

Motion estimation in the H.264 encoder is one of the most computationally intensive sub-kernels. Motion estimation is performed on each macroblock of the incoming frame, where the Sum of Absolute Differences (SAD) is used to compare the current macroblock with the reference macroblocks to find the best possible match. The number of SADs that need to be computed for a macroblock heavily depends on the motion contained in that particular macroblock. A macroblock containing fast moving objects will require more SADs compared to a macroblock of slow moving objects. Figure 6.1 shows the number of SADs that were computed for the first 200 macroblocks of the second frame (the first frame does not require motion estimation) of the 'pedestrian' video sequence [1]. It is obvious that the workload of the motion estimation sub-kernel varies significantly at run-time; the number of computed SADs

H. Javaid and S. Parameswaran, *Pipelined Multiprocessor System-on-Chip for Multimedia*, DOI: 10.1007/978-3-319-01113-4_6,

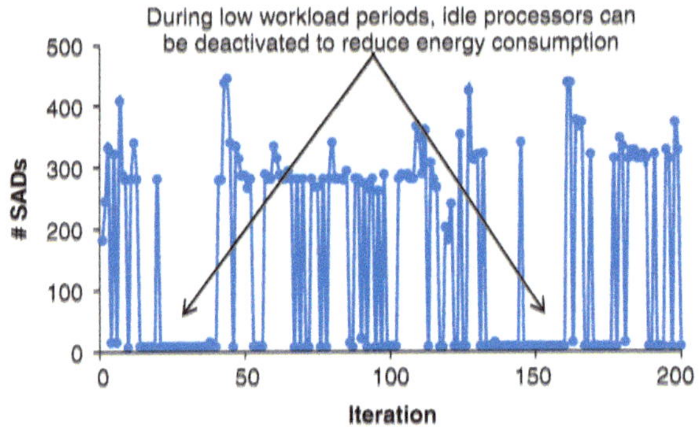

Fig. 6.1 Number of SADs computed during different iterations of the motion estimation sub-kernel

can go as high as 500 and as low as 10 with an average and standard deviation of 154 and 153 SADs respectively.

Consider the motion estimation stage of a worst-case pipelined MPSoC which contains 17 processors in parallel to process HD720p video. In addition, consider that each processor can compute 30 SADs within the throughput constraint. Thus, in total the motion estimation stage is capable of computing $17 \times 30 = 510$ SADs which is enough to sustain throughput at all times (worst case is 500 SADs). During low workload periods (marked in Fig. 6.1), only one processor is doing useful work because these periods require computation of less than 30 SADs (which can be handled by a single processor). Thus, during the marked low workload periods, the other 16 processors will be idle, resulting in inefficient utilisation of resources and increased energy consumption of the pipelined MPSoC. In contrast, in an adaptive pipelined MPSoC, a *resource-aware approach* would have shared the idle processors with other stages at run-time, while an *energy-aware approach* would have deactivated the idle processors at run-time to reduce energy consumption.

In summary, design-time balanced, worst-case pipelined MPSoCs do provide high performance but at the cost of inefficient resource utilisation and increased energy consumption. Hence, worst-case pipelined MPSoCs do not provide a resource- or energy-aware platform for advanced multimedia applications such as H.264/AVC [2], AVS [3], VC1 [4] which exhibit huge variations in their workload at run-time due to the adaptive nature of their algorithms and input data. As a result, applicability of worst-case pipelined MPSoCs as a platform for multimedia applications in portable devices is limited because of the area footprint and energy constraints in such devices.

In this chapter, a worst-case pipelined MPSoC is augmented with a run-time management (balancing) technique so that it can adapt itself to run-time varying workloads. To this end, an adaptive pipelined MPSoC architecture is proposed where stages with significant run-time variations in workload are implemented using *Main Processors* and *Auxiliary Processors*. The main processor uses differing number of auxiliary processors considering run-time workload variations. The run-time management technique uses a combination of the application's execution and knowledge (algorithmic and data properties) to predict the upcoming workload (number of auxiliary processors for a main processor). To reduce energy consumption of the adaptive pipelined MPSoC, the idle auxiliary processors are either clock-gated or power-gated.

The works in [5–7] considered adaptability in pipelined MPSoCs. Guo et al. [7] proposed a dynamic voltage scaling approach to reduce the voltage to processors with low workload, while [5, 6] showed the application of Dynamic Voltage and Frequency Scaling (DVFS) in pipelined MPSoCs. All these works used a feedback controller to monitor the occupancy level of the queues to determine when to increase or decrease the frequency-voltage levels of a processor. Thus, these works used the execution history of the application and were reactive in nature. The run-time management techniques proposed in this chapter not only utilise the application's execution history, but also the application's knowledge to proactively predict the upcoming workload. An application's knowledge should be used in workload prediction because an application knows (or may know) by far the most about its future workload [8]. However, unlike [8] where just algorithmic properties in a uniprocessor system were employed, more diverse application knowledge (algorithmic and data properties) are considered in a pipelined MPSoC in this chapter.

From a practical perspective, the provision of DVFS circuitry for MPSoCs with more than two processors is very expensive [9]. Furthermore, the large overhead of DVFS control circuitry limits its use to systems requiring only coarse-grained run-time management [10]. The shrinkage of the dynamic range of voltage-frequency operational points due to downward scaling of supply voltage has also limited DVFS use, and has given rise to the use of clock-gating, power-gating and multiple power states. Therefore, the adaptive pipelined MPSoC architecture proposed in this chapter allows a main processor to manage its auxiliary processors by either clock- or power-gating them. Chapter 7 extends this work for multiple power states.

6.2 Adaptive Pipelined MPSoC Architecture

Figure 6.2 shows a typical pipelined MPSoC, comprised of various pipeline stages. Adaptability is introduced in such a pipelined MPSoC by the use of *Main Processors (MPs)* and *Auxiliary Processors (APs)*. Thus, each processor in the pipelined MPSoC is either an MP or an AP. A processor is categorised as an MP if its sub-kernel(s) is (are) executed for every iteration of the multimedia application, that is it is always active. On the other hand, an AP is a processor whose mapped sub-kernel(s) will be executed for a maximum of the total number of the application's iterations, that is, it can be idle during some iterations. Adaptability in stages with significant run-time variation in their workloads is realised by implementing those stages using a combination of MPs and APs, where a pool of APs is connected to an MP using FIFOs. In addition, MPs and their APs can have access to a shared memory if common data needs to be shared between them.

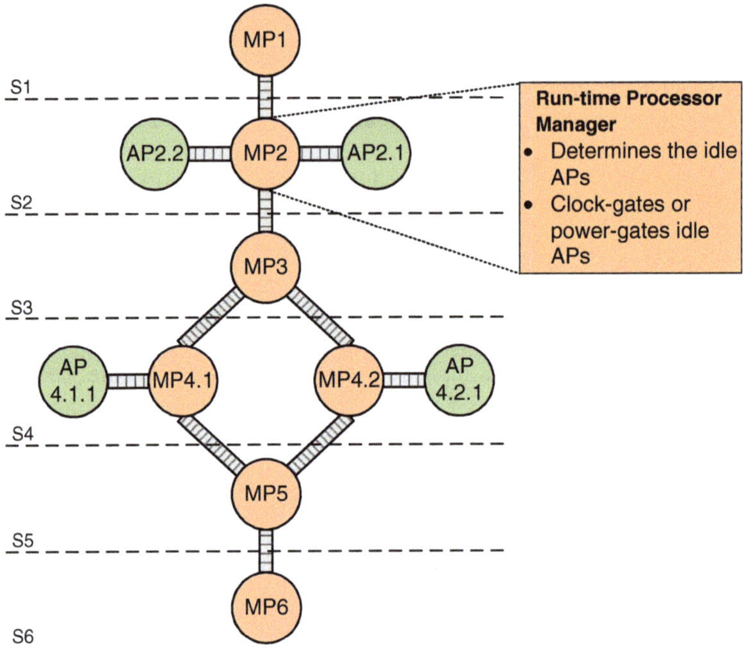

Fig. 6.2 Adaptive pipelined MPSoC's architecture

An example of an adaptable stage is S4 in Fig. 6.2 which contains two MPs (MP4.1 and MP4.2) that will be active at all times. These MPs will use their corresponding APs (AP4.1.1 and AP4.2.1) only when the workload increases beyond their capacities. In other words, MPs handle the nominal workload while APs handle the extra workload by working in parallel with their corresponding MPs. If all the APs of an adaptive pipelined MPSoC are considered to be MPs, then it will become a worst-case pipelined MPSoC, where all the processors will be always active (thus only the existence of MPs). It should also be noted that stages with almost constant workload do not need APs and are only implemented with MPs; for example, stage S3 in Fig. 6.2. Thus, an adaptive pipelined MPSoC provides an effective implementation platform for advanced multimedia applications which contain stages with both almost constant workload (such as DCT) and run-time varying workload (such as motion estimation).

The proposed adaptive pipelined MPSoC is a hybrid system due to the co-existence of MPs and APs, and its adaptability can be exploited in several ways. For example, a resource-aware run-time manager could be deployed to allocate the idle APs of one stage to another stage that is currently experiencing high workload, resulting in efficient resource utilisation.[1] Another example is to deploy an energy-aware run-time manager where the APs are deactivated during idle iterations to reduce the energy consumption. This chapter focuses on the later by proposing a run-time processor manager, considering the support for clock- and power-gating (two well-known power reduction techniques) based deactivation of idle APs.

The architecture of the adaptive pipelined MPSoC allows for both a centralised and a distributed run-time manager. However, a distributed processor manager is proposed in this chapter where an MP adapts to its varying workload by activating/deactivating its APs, independent of other MPs. Such a distributed approach has the advantage of scalability over a centralised processor manager. Furthermore, each stage can tweak the run-time management heuristics (see Sect. 6.6) according to its own workload profile. Therefore, highly customised, per-stage run-time managers can be deployed in the adaptive pipelined MPSoC to lower their performance and energy overheads.

Figure 6.2 zooms in on one of the MPs to illustrate that each MP with a pool of APs has a run-time processor manager. This processor manager determines the idle APs of an MP by considering the workload at run-time for every iteration of the application. For example, if the run-time manager determines AP14 and AP15 to be idle for an MP with 16 APs, then AP14 and AP15 will either be clock- or power-gated to reduce energy consumption. In this chapter, either the idle AP is only clock-gated or only power-gated without the provision for selective use of clock- and power-gating. Chapter 7 will extend this work for selective use of different power reduction techniques through the use of multiple power states.

6.3 A Design Flow

The design flow to create an adaptive pipelined MPSoC is shown in Fig. 6.3. The sub-kernels of a multimedia application are mapped to the processors of a pipelined MPSoC. The pipelined MPSoC is then passed through a customisation phase, where each processor is customised according to the sub-kernel(s) mapped on it to balance the stages of the pipelined MPSoC. Further details can be found in Chaps. 3 and 5. The customised pipelined MPSoC is created using worst-case parameters so that it can deliver the required throughput at all times.

A worst-case pipelined MPSoC is transformed into an adaptive pipelined MPSoC by the addition of a run-time manager. Firstly, the worst-case pipelined MPSoC is profiled with various data inputs to gather statistical information such as minimum, maximum and average workload. For example, in the motivational example of Sect. 6.1 where 17 processors were used in the motion estimation stage, a minimum and a maximum of 10 and 500 SADs are computed respectively with a standard

[1] Resource sharing would require connection of an AP with multiple MPs. Since resource sharing is not considered in this monograph, each AP is connected to only one MP, though multiple APs can be connected to a single MP.

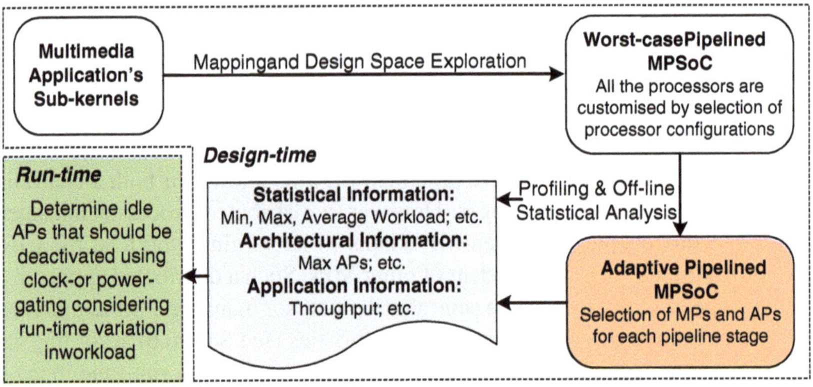

Fig. 6.3 A design flow for adaptive pipelined MPSoCs

deviation of 153 SADs. In addition, the throughput constraint of the multimedia application is used to compute the period of each pipeline stage (that is, the maximum number of clock cycles for an iteration), and is referred to as T_C. The profiling information also records the amount of workload a processor can handle within T_C clock cycles. For example, a processor in motion estimation stage can compute 30 SADs in T_C clock cycles. Using the profiling and statistical information, the number of MPs and APs is decided for each stage of the worst-case pipelined MPSoC. For the running example, motion estimation stage can be implemented with one MP and 16 APs because the minimum workload of 10 SADs can be handled by one processor. A similar procedure is also used for the other stages of the pipelined MPSoC where run-time adaptation is required. In summary, statistical information from the profiling is used to decide the number of MPs and APs for each stage of the adaptive pipelined MPSoC. Then, the information gathered off-line (statistical, architectural and application) is used by the run-time manager in addition to run-time monitoring of the workload to activate and deactivate APs at run-time.

6.4 Problem Statement

Given an adaptive pipelined MPSoC and the off-line gathered information, the goal is to determine "when" and "how many" APs to activate and deactivate at run-time for each MP under run-time variations in workload so that the required throughput is delivered with minimal degradation and maximal reduction in energy consumption.

The challenge is to predict the correct number of APs for an iteration because the use of an incorrect number of APs will either result in loss of throughput (when less than required APs will be used) or an increase in energy consumption (when more than the required number of APs will be used which could otherwise have been deactivated). A feedback based approach, particularly the one that is only based on the application's execution history, suffers from slow response because the run-time manager cannot detect workload variation until the current iteration has finished, and thus may result in significant loss of throughput (see Sect. 6.7.2). In other words, feedback based approaches are reactive in nature rather than proactive. Additionally, multiple activations/deactivations of an AP within the same iteration may lead to an increase in energy consumption rather than its reduction due to the overhead of activation and deactivation. Thus, a sophisticated run-time manager is required to decide the number of APs that should be activated for an MP, considering more than just the application's execution history. Furthermore, such a run-time manager should have low performance and energy overheads.

The next two sections explain how application knowledge can be leveraged to predict the workload of (some of the stages of) a multimedia application, and how that prediction, in addition to the application's execution history, can be used by an MP to manage its APs.

6.5 Leveraging Application Knowledge

In multimedia applications, much information is available from the application and input data, such as texture, brightness, size and homogeneity of the macroblocks or frames in an H.264 video encoder/decoder [11]. Typically, a pre-processing stage is employed in multimedia applications to analyse such information [11]. The pre-processing stage processes the input data to extract useful information for the video processing system in advance for run-time adaptation [12]. In this chapter, the author uses such information at system-level in the run-time processor manager to decide the number of APs for an MP.

6.5.1 An H.264 Video Encoder Example

In this section, one piece of information available at the pre-processing stage to the motion estimation sub-kernel in an H.264 video encoder is elaborated. Consider the pre-processing stage categorises the macroblocks of a frame as either low or high motion macroblocks. Low motion macroblocks typically contain slow moving objects and are homogeneous while high motion macroblocks are textured and contain fast moving objects. Depending on the texture/variance of the current macroblock (mb_i) and the predicted SADs of the neighbouring macroblocks, the workload in number of SADs for the current macroblock can be predicted as follows:

if $Var(mb_i) < Var_{th}$ *and* $SAD(mb_i) < SAD_{th}$ **then**
$mb_i is\ a\ low\ motion\ macroblock$
else
$mb_i\ is\ a\ high\ motion\ macroblock$

where

$$Var(mb_i) = \frac{1}{N}\sum_{j=1}^{N}(P_j - AvgBrightness(mb_i))^2$$

$$AvgBrightness(mb_i) = \frac{1}{N}\sum_{j=1}^{N}P_j + N/2$$

$$SAD(mb_i) = \text{median}\{SAD(mb_{right}), SAD(mb_{top}), SAD(mb_{topRight})\}$$

The Var_{th} and SAD_{th} are threshold values for the variance and number of SADs, and are typically obtained through regression analysis [11]. The term P_j refers to the jth pixel of a macroblock, and N is the total number of pixels.

The number of SADs for low- and high-motion macroblocks are obtained through the Probability Density Function (PDF), which are shown for two test video sequences ("station" and "tractor") in Fig. 6.4. Based on the category of the current macroblock (low- or high-motion), the correct distribution is used to obtain the zone of high probability (a probability of 84 % for Gaussian distributions) [13]. For example, the two vertical (dotted-blue) lines in the "station" graph is the range for the number of SADs for low motion macroblocks, while the range represented by the two vertical (dotted-red) lines in the "tractor" graph is for the high-motion macroblocks. From Fig. 6.4, the range for low- and high-motion macroblocks will be [0, 150] and [150, 600] number of SADs. It should be noted that the prediction is fuzzy as it is in the form of a range.

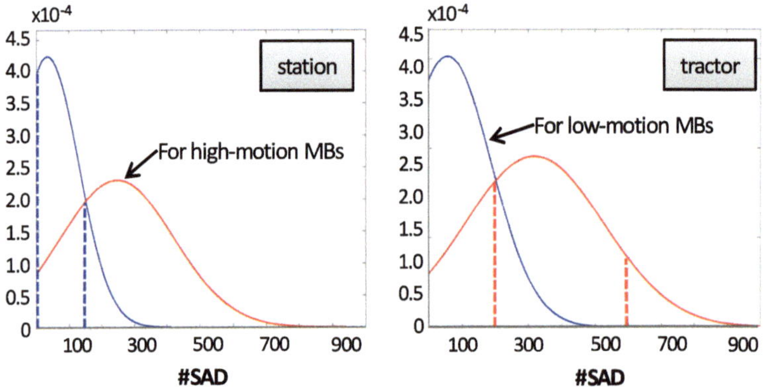

Fig. 6.4 Probability density function of number of SADs for low- and high-motion macroblocks (MBs)

For each range of the workload, an off-line analysis is performed to obtain the number of processors that can handle that much workload. For example, the ranges of [0, 150] and [150, 600] number of SADs can be converted to [0, 5] and [6, 20] number of processors respectively if the off-line analysis revealed that each processor can handle 30 SADs within one iteration.

In summary, the workload ranges are computed off-line and stored in the pre-processing stage in the form of a lookup table to reduce the run-time overhead. At run-time, the pre-processing stage first categorises the current macroblock (by analysing its variance and number of SADs of neighbouring macroblocks), and then uses its category to obtain the corresponding workload range from the lookup table. Hence, the pre-processing stage can predict the workload of current iteration (an iteration processes a macroblock) in the form of a range (in number of processors) for the motion estimation stage of the H.264 video encoder.

6.6 Processor Management Heuristics

This section describes two heuristics to manage APs at run-time: the first heuristic is based on only the application's execution history, while the second heuristic is based on both the application's execution and knowledge. For the sake of simplicity, the heuristics are explained from the perspective of one MP; however they are equally applicable to other MPs of the adaptive pipelined MPSoC. The heuristics exploit the fact that an MP cannot exceed T_c clock cycles (throughput constraint) during an iteration in order to guarantee the required throughput. Therefore, at some time instants during the current iteration, the heuristics check whether there is a possibility of violating the T_c constraint. Then, according to the predicted workload (based on either the application's execution or a combination of the application's execution and knowledge) APs are either activated or deactivated. The following terms are introduced to explain the heuristics:

- $W_a[i]$: Actual workload of the ith iteration, equal to the number of APs that are active at the end of the ith iteration. For the current iteration, $W_a[i]$ holds the number of currently active APs.
- $W_p[i]$: Predicted workload for the ith iteration in number of required APs from the pre-processing stage.
- $CC[i]$: Clock cycles spent by MP in its ith iteration which are monitored at run-time.
- AP_M: The total number of APs for an MP where APs are denoted as AP_0, AP_1, ... AP_{M-1}.
- *obsW* (*observation window*): The number of consecutive, previous iterations used at run-time for observation of the application's execution.
- *calW* (*calculation window*): The number of consecutive, previous iterations used at run-time for calculation of the average workload in those iterations. The *calW*

Function isLowWorkload

```
// Called during k-th iteration to check whether it will have
low workload or not based on previous iterations in the
observation window (obsW)
for o=0; o<obsW; o++ do
    if CC[k-1-o] > Tc/2 then
        return not low workload iteration
return low workload iteration
```

Function isHighWorkload

```
// Called during k-th iteration to check whether it will have
high workload or not based on previous iterations in the
observation window (obsW)
for o=0; o<obsW; o++ do
    if CC[k-1-o] > Tc then
        return high workload iteration
return not high workload iteration
```

is restricted to be less than or equal to $obsW$ which means that the calculation window is a subset of observation window.

The heuristics use *isLowWorkload* and *isHighWorkload* functions to determine whether the current iteration will have low workload or high workload, based upon the application's execution history. If the clock cycles ($CC[i]$) of any one of the previous iterations in the observation window ($obsW$) exceeded T_c, then the current iteration is considered to be a high workload iteration. This is because the violation of throughput constraint in the near past suggests that the chances of exceeding T_c during current iteration are high. If the clock cycles of all the previous iterations in the observation window were less than $T_c/2$, then the current iteration is considered to be a low workload iteration. This is because low workload iterations in the near past suggest that there are more chances of current iteration being a low workload iteration.

6.6.1 Application Execution Based Heuristic (Exe Heuristic)

The Exe heuristic monitors the workload of previous iterations (in observation and calculation windows) to keep a record of the average workload (in number of APs) of those iterations. The average workload is then used as the predicted workload of the current iteration, that is, the number of APs that will be required during the current iteration. The heuristic is shown in Algorithm 3.

Function getAPsFromExecution

// Called during the k-th iteration to obtain the average number of active APs from previous iterations in the calculation window (calW)

1 APsFromExecution = 0;
2 **for** *c=0; c<calW; c++* **do**
3 APsFromExecution += $\lfloor \frac{(CC[k-1-c]\times(W_a[k-1-c]+1))}{T_c} \rfloor$;
4 **return** APsFromExecution/calW;

The *getAPsFromExecution* function computes the average number of APs that should have been active in the previous iterations where the number of the iterations to consider is equal to the length of the calculation window ($calW$). The factor $(CC[k-1-c]\times(W_a[k-1-c]+1))$ computes the workload of the $(k-1-c)$th iteration in clock cycles where the addition of one in $W_a[k-1-c]$ is due to the presence of the MP. The number of APs that should have been active in $(k-1-c)$th iteration are then calculated by dividing the workload in clock cycles by T_c. For example, if the last iteration used 3,000 clock cycles ($CC[k-1] = 3,000$) and three APs were active ($W_a[k-1] = 3$) under a throughput constraint of 9,000 clock cycles ($T_c = 9,000$) and calculation window of one iteration ($calW = 1$), then the average workload in numbers of APs is one $\left(\lfloor \frac{3,000\times(3+1)}{9,000} \rfloor = 1\right)$. Note that $W_a[k-1-c]$ alone is not used because it is not the true workload of an iteration. For example, in the running example, three APs were active in the last iteration; however, only one of them should have been active which is the true workload of the last iteration. At the end (line 4), *getAPsFromExecution* function returns the average workload of the iterations in the calculation window in the number of APs.

The Exe heuristic uses the output of the *getAPsFromExecution* function as the predicted workload for the current iteration (line 1). Afterwards, it checks whether

Algorithm 3: Exe Heuristic (for the sake of simplicity, boundary cases are not reported here)

// Called at the start of k-th iteration to decide the number of APs

1 $W_p[k]$ = getAPsFromExecution();
2 **if** *isHighWorkload()* **then**
3   **if** $W_p[k] > W_a[k]$ **then**
4   addAPs = $W_p[k] - W_a[k]$;
5   ACTIVATE *addAPs* many more APs
6 **if** *isLowWorkload()* **then**
7   **if** $W_p[k] < W_a[k]$ **then**
8   subAPs = $W_a[k] - W_p[k]$;
9   DEACTIVATE *subAPs* many APs

the current iteration will have a high (line 2) or low (line 7) workload based on the clock cycles of the previous iterations in the observation window using *isHighWorkload* and *isLowWorkload* functions respectively. If the current iteration is considered to be a high workload iteration and the predicted workload is higher than the current workload (line 3) then *addAPs* (equal to the difference between predicted and current workloads) many extra APs are activated (lines 4–5). On the other hand, if the current iteration is a low workload iteration and the predicted workload is less than the current workload (line 7), then *subAPs* (equal to the difference between predicted and current workloads) many APs are deactivated (lines 8–9). Note that boundary cases to ensure the number of active APs does not exceed AP_M and some optimisation steps are skipped for the sake of simplicity.

The Exe heuristic keeps the minimum amount of information so that its run-time overhead is low. Furthermore, the average workload for a given iteration is updated at run-time based on the execution history of the calculation window. However, the Exe heuristic will have a slow response during sudden changes in workload, resulting in significant loss of throughput (see Sect. 6.7.2). The Exe heuristic portrays typical feedback controller based techniques that have been used in earlier works [5–7], and hence is a representative of those techniques in the adaptive pipelined MPSoC proposed here.

6.6.2 Application Knowledge Based Heuristic (Know Heuristic)

As explained in Sect. 6.5, a pre-processing stage is available which can predict the workload range of each iteration in the number of APs using the application knowledge. The Know heuristic combines such prediction with statistical information gathered off-line and the application's execution monitored at run-time to better manage APs with quick response. The following terms are used in addition to the ones described in Sect. 6.6:

- W_{SD}: Standard deviation of the MP's workload in the number of APs, available from off-line statistical analysis.
- AP_T: Minimum number of APs that should be activated or deactivated at an instant during the current iteration, which is computed off-line. The value of AP_T affects the response time of the MP, that is, how quickly an MP adapts to the variation in its workload. For example, a high value of AP_T will enable a quick response by activating a large number of APs, reducing the impact on the throughput. However, a very high value (close to AP_M) will result in most of the APs being active at all times, reducing the amount of energy reduction. Therefore, the author computes AP_T such that the MP can respond to a variation of $\frac{W_{SD}}{2}$ (half of the workload's standard deviation) within T_c clock cycles to allow a reasonable trade-off between throughput degradation and energy reduction. Consider that the APs are activated when the current iteration's clock cycles have reached $T_c/2$ and $3T_c/4$ (which will be further explained later), then:

$$\frac{W_{SD}}{2} = \frac{AP_T}{2} + \frac{AP_T}{4}$$
$$AP_T = \frac{2W_{SD}}{3}$$

The factors $\frac{AP_T}{2}$ and $\frac{AP_T}{4}$ refer to the workload that can be distributed to APs at $T_c/2$ and $3T_c/4$ time instants respectively. Thus, if $AP_T = \frac{2W_{SD}}{3}$ many APs are activated at $T_c/2$ and $3T_c/4$ time instants, then $\frac{W_{SD}}{2}$ workload can be handled by those APs without exceeding T_c clock cycles. Further variations in the workload are addressed by the workload predictions from the application's knowledge.

- *getAPsFromKnowledge*: This function returns the minimum of the workload range predicted for the current iteration by the pre-processing stage. Recall from Sect. 6.5 that the pre-processing stage categorised each macroblock as either low- or high-motion macroblock and the corresponding workload ranges in the number of APs were [0, 5] and [6, 20], then *getAPsFromKnowledge* will return 0 and 6 for low- and high-motion macroblocks (iterations) respectively.
- AP_D: The maximum difference between the minimums of consecutive workload ranges. For example, $AP_D = 6$ for the two workload ranges of [0, 5] and [6, 20].

The Know heuristics has two parts which are triggered at different time instants during the current iteration, which is shown in Algorithm 4. The first part (lines 1–11) is triggered at the start of each iteration to decide the number of APs in advance to maximally minimise the penalty on throughput. The minimum number of APs predicted for the current iteration are obtained using the *getAPsFromKnowledge* function (line 1). If the predicted workload is more than the current workload, then *addAPs* many extra APs are activated (lines 2–4). On the other hand, if the predicted workload is less than the current workload, then some or all the APs are deactivated. The number of APs to deactivate are computed by utilising not only the predicted workload, but also the execution history and off-line statistical information. Thus, the first step is to check whether the current iteration is considered a low workload iteration or not based on execution history (line 6). If so, then the predicted workload is adjusted by taking the maximum among the application's knowledge and execution based predictions (line 7). If the adjusted predicted workload is less than the current workload (which means that both the application's knowledge and execution history suggested use of less number of APs), then *subAPs* many APs are deactivated (lines 8–11). The minimum operation in line 10 ensures that not more than AP_T many APs are deactivated so that the MP can respond back quickly when there is a sudden increase in the workload, incorporating the information from the off-line statistical analysis.

Although the first part of the Know heuristic activates or deactivates APs in advance for minimum degradation of throughput and maximum reduction of energy, the throughput constraint can still be violated because both the application's knowledge and execution based predictions are fuzzy. Therefore, the second part of the Know heuristic is triggered at $T_c/2$, $3T_c/4$ and T_c instants to check the possibility of violating the throughput constraint. When triggered at $T_c/2$ and $3T_c/4$ instants

Algorithm 4: Know Heuristic (for the sake of simplicity, boundary cases are not reported here)

```
// Called at the start of k-th iteration to decide the number of APs
1  W_p[k] = getAPsFromKnowledge();
2  if W_p[k] > W_a[k] then
3      addAPs = W_p[k] − W_a[k];
4      ACTIVATE addAPs many more APs
5  else
6      if isLowWorkload() then
7          W_p[k] = max { W_p[k], getAPsFromExecution() };
8          if W_p[k] < W_a[k] then
9              subAPs = W_a[k] − W_p[k];
10             subAPs = min { subAPs, AP_T };
11             DEACTIVATE subAPs many APs

// Called at T_c/2, 3T_c/4 and T_c time instants during k-th iteration to activate more APs
12 if CC[k] == T_c/2 || CC[k] == 3T_c/4 then
13     if isHighWorkload() then
14         addAPs = getAPsFromExecution() − W_a[k];
15         addAPs = max { addAPs, AP_T };
16         ACTIVATE addAPs many more APs
17 else if CC[k] == T_c then
18     addAPs = AP_D;
19     ACTIVATE addAPs many more APs
```

(lines 12–16), it checks whether the current iteration is considered a high workload iteration or not. If so, then *addAP* many extra APs are activated. The value of *addAPs* is computed from the application's execution history and AP_T (lines 14–15) because the application's knowledge based prediction has already been utilised at the start of the current iteration. It is possible that the variation in workload is too high and even after previous activations of the APs, the clock cycles of current iteration have reached T_c. At this instant, AP_D many extra APs are activated to ensure that the remaining workload is handled within the next T_c clock cycles, introducing a worst-case penalty of T_c clock cycles. This is because if, for example, the current iteration is predicted to have a [0, 6] workload range, then it could not have required activation of more than $AP_D = 6$ APs. Otherwise, the current iteration would have been categorised to have a [6, 20] workload range by the pre-processing stage (considering there are only two categorisations as described in Sect. 6.5). Note that boundary cases, to ensure number of active APs does not exceed AP_M, and some optimisation steps are skipped for the sake of simplicity.

The Know heuristic ensures that the APs are not activated and deactivated multiple times within the current iteration which otherwise would incur significant overhead of activation and deactivation. The first part of the heuristic only deactivates APs if the current iteration is considered a low workload iteration (line 6) which is mutually exclusive to the activation condition in the second part (line 13), avoiding unnecessary activation and deactivation of APs. Furthermore, the deactivated APs will remain deactivated until T_c clock cycles in the current iteration.

These run-time processor management heuristics are executed on MPs with a pool of APs; however, the values of W_{SD}, AP_M, AP_T, AP_D, $obsW$ and $calW$ will vary from one MP to another depending upon their workload profiles. Note that the length of the observation and calculation windows ($obsW$ and $calW$) will affect the outcome of the heuristics; however, the reason for the use of variable window lengths is that a designer can tweak the heuristics for different stages of the adaptive pipelined MPSoC based on their workload profiles. The proposed run-time management heuristics do not use any complex computations and hence their overhead is small (see Sect. 6.7.2).

6.6.3 System-Level Overview

The system-level implementation of the proposed adaptive pipelined MPSoC with the processor manager, executing a multimedia application such as H.264 video encoder, is shown in Fig. 6.5. A multimedia application is implemented as a combination of pre-processing and multimedia systems. The pre-processing system extracts the features of incoming frames to provide useful information to the multimedia system for run-time adaptations. For example, a pre-processing stage can categorise macroblocks according to the motion contained in them as described in Sect. 6.5. The multimedia system implements the video codec on an adaptive pipelined MPSoC. A processor manager is implemented for each of the MPs with a pool of APs. More specifically, the processor manager uses either the Exe heuristic or the Know heuristic to deactivate the idle APs at run-time.

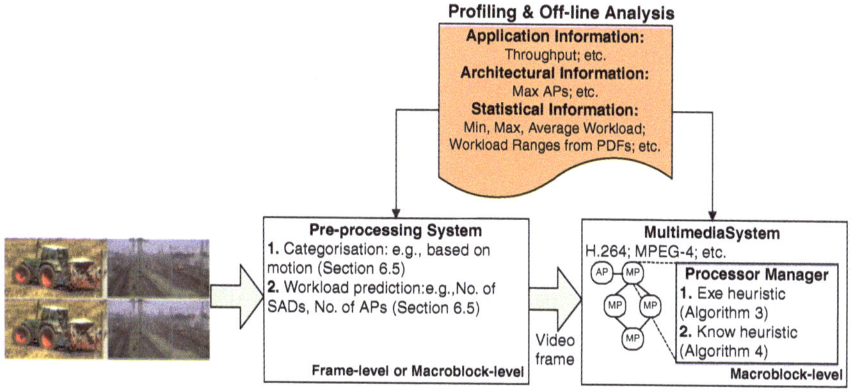

Fig. 6.5 A system-level implementation overview of adaptive pipelined MPSoCs

Statistical, architectural and application information obtained through profiling and off-line statistical analysis is used to guide the two systems at run-time. For example, the workload ranges (number of SADs for the motion estimation stage) are obtained through the statistical analysis which are then converted to equivalent number of APs through profiling for workload prediction at run-time. Other information such as the minimum, average and maximum workload is also provided. The pre-processing system is expected to work at either the frame-level or macroblock-level so that the workload predictions for all the macroblocks of a frame is available to the multimedia system which is working at the macroblock-level. Thus, the proposed run-time manager is applicable to all advanced macroblock based video coding applications such as H.264, MPEG-4, AVS, and VC1.

The proposed adaptive pipelined MPSoC and processor manager is applicable to all multimedia applications where a pre-processing stage can be deployed to guide the run-time manager. If a pre-processing system is not available, then the processor manager can use the application knowledge from the multimedia system (for example, the actual number of SADs of the previous macroblocks) to predict the future workload; however, such a prediction would be less accurate. The variables in the run-time management heuristics (W_{SD}, AP_M, AP_T, etc.) allow them to be tweaked according to the workload profiles of a sub-kernel or stage of a multimedia application. It should be noted that the run-time processor manager proposed here can be used in architectures other than the pipelined MPSoCs. For example, in a master-slave architecture, the master processor will execute the processor management heuristics to deactivate the idle slave processors.

6.7 HD720p H.264 Video Encoder Case Study

Implementation of an H.264 video encoder for HD720p resolution at 30 fps on an adaptive pipelined MPSoC is presented in this section for comparison and evaluation of the proposed heuristics.

6.7.1 Implementation Details

The adaptive pipelined MPSoC for H.264 video encoder was implemented using Xtensa LX3 [14] family of processors, which come with the RC-2010.1 tool suite. Like Chap. 4, the processors were customised automatically by utilising the XPRES tool and the adaptive pipelined MPSoC was created in the XTMP environment. The XTMP uses the XT-XENERGY tool to measure the power and energy of the processors in a multiprocessor environment. Hence, the author obtained the throughput and energy of the adaptive pipelined MPSoC from the XTMP, where all the processors were running at 1 GHz and XT-XENERGY was configured for a given 45 nm technology.

The H.264 video encoder application graph from Chap. 3 is reproduced in Fig. 6.6 with additional details. Due to the feedback loop between the Loop Filter (LF) and the Motion Estimation (ME) sub-kernels, execution of this task graph at frame-level will introduce unacceptable delay between each iteration, and thus will provide no useful benefit. Thus, the task graph is executed at macroblock-level where each sub-kernel processes one macroblock in an iteration (which is typical of real-time implementations of the H.264 encoder/decoder [15]). Furthermore, the entropy coding processes macroblocks in parallel to the reconstruction path (ITQ and LF) to increase the throughput of the system. The annotations around the arrows show the amount of data (buffer sizes in adaptive pipelined MPSoC) in bytes being transferred in each iteration. For example, the CC sub-kernel sends the Y component of a 16×16 macroblock to ME sub-kernel, which is a transfer of 256 bytes in each iteration. It should be noted that the CC and IP/MC sub-kernels send data to IP/MC and ITQ sub-kernels respectively in advance (bypassing the intermediate sub-kernels) to increase the throughput of the system. In this task graph, ME, IP/MC and EC sub-kernels exhibit run-time variation in their workloads, requiring a run-time processor manager for each of them. However, in this case study, the author only deployed adaptability for the ME sub-kernel, providing a proof of concept for the proposed run-time management heuristics. Thus, all the sub-kernels in Fig. 6.6 were mapped on MPs in the adaptive pipelined MPSoC, except the ME stage where a combination of MPs and APs was used.

An H.264 encoder supporting HD720p at 30 fps needs to process $30 \times 3600 = 108{,}000$ macroblocks/s. Since processors are running at 1 GHz, a macroblock should be processed within $\frac{1 \times 10^9}{108{,}000} \approx 9{,}260$ clock cycles. Thus, $T_c = 9,100$ clock cycles is used to have a conservative throughput constraint. The profiling and off-line statistical

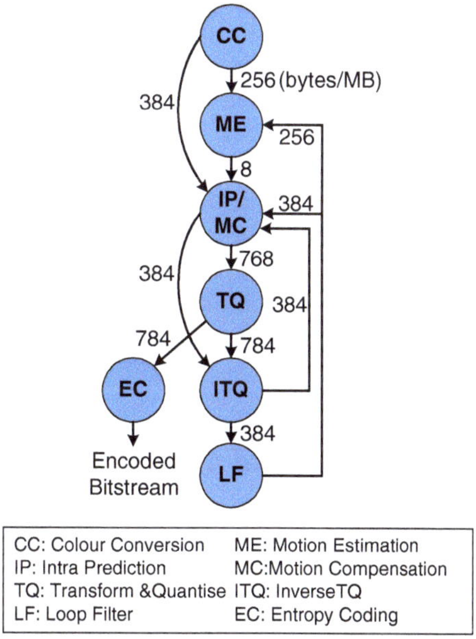

Fig. 6.6 Details of an H.264 video encoder application

analysis of the ME sub-kernel (with a fast motion estimator [16]) using various input video sequences yielded 225 average number of SADs with a maximum of 500 SADs and an average standard deviation of 200 SADs per macroblock. Furthermore, the ME processor was able to compute only 30 SADs in T_c clock cycles. The pre-processing stage provided workload prediction by categorising macroblocks as either low-, medium- or high-motion macroblocks [12]. The workload range of each category in the number of APs was computed using off-line analysis and was saved in a lookup table for use at run-time. These ranges were [0, 4], [5, 10] and [11, 16] (number of APs) for low-, medium- and high-motion macroblocks respectively. Using the above described information and setup:

- $\mathrm{W}_{SD} = \frac{200}{30} = 6.67$.
- $\mathrm{AP}_M = \lfloor\frac{500}{30}\rfloor = 16$ (17 processors including the MP).
- $\mathrm{AP}_T = \frac{2W_{SD}}{3} = \frac{2\times 6.67}{3} \approx 4$.
- $\mathrm{AP}_D = 6$.

Therefore, sixteen APs were connected to the MP in the ME stage of the adaptive pipelined MPSoC. These APs could be either clock- or power-gated when idle. The author assumed no overhead for clock-gating an AP as it can be done in a few clock cycles. However, for power-gating an AP, an activation/deactivation time and energy consumption of 100 ns (100 clock cycles at 1 GHz) and 250 nJ were assumed respectively, which are typical of processor-level power-gating [17, 18]. Note that an AP is activated and then appropriate data is sent to it by the MP to overlap the activation time with the communication time as the FIFOs between the MP and APs are always active. The communication latency of sending data (at least 256 ns assuming a byte transfer takes at least 1 clock cycle @ 1 GHz) to APs after activating them is larger than the activation overhead of power-gating (100 ns) and hence did not affect the throughput of the pipelined MPSoC.

The MP monitored its execution in clock cycles per iteration using a built-in timer module. The MP and APs executed the same code of ME sub-kernel except that the MP also executed the two heuristics. The Exe heuristic used $obsW = 8$ and $calW = 1$, while the Know heuristic used $obsW = 2$ and $calW = 1$. This is because the Exe heuristic needs longer windows of the application's execution to better capture the run-time variation in workload compared to the Know heuristic, where the application's knowledge compensates for the error in the workload profile captured from smaller windows of execution.

6.7.2 Results and Analyses

The adaptive pipelined MPSoC was executed for five different HD720p (high definition) video sequences: pedestrian; sky; station; sunflower; and, tractor [1]. The hardware-related details of the adaptive pipelined MPSoC are reported in Table 6.1, where the second and third columns summarise the area and power consumption

Table 6.1 Hardware-related details of the adaptive pipelined MPSoC for 'pedestrian' video sequence

Processor	Area (KGates)	Power (mW)	
		Dynamic	Leakage
CC	92.44	43.24	6.80
ME	103.23	41.49	8.40
ME-AP0–ME-AP15	103.23	≈29.00	≈6.51
IP/MC	103.65	40.37	7.69
TQ	91.56	48.42	6.50
ITQ	93.50	49.68	7.07
LF	87.76	41.68	6.48
EC	90.12	44.75	6.49

of the MPs and APs for the 'pedestrian' video sequence. Other video sequences exhibited similar trends, and thus are not reported here.

Figure 6.7a illustrates the adaptability of the ME stage at the iteration level for the 'pedestrian' video sequence. The figure plots the true workload (number of SADs computed in each iteration/30 because each processor can compute 30 SADs in

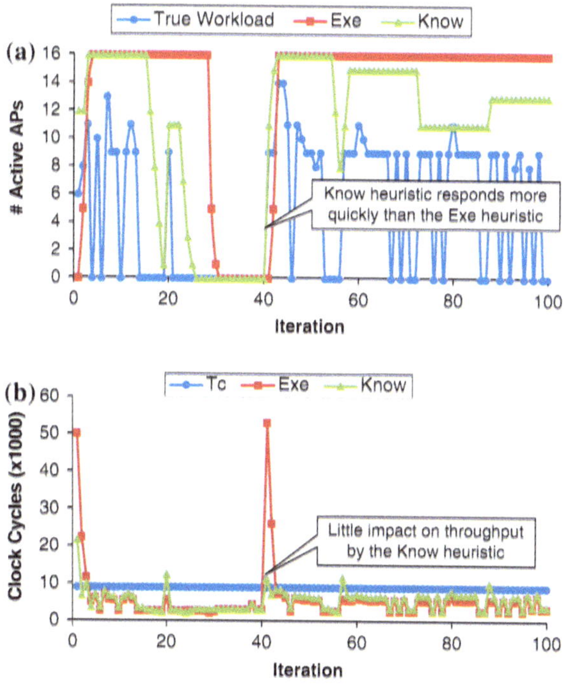

Fig. 6.7 **a** Adaptability and **b** Throughput for the 'pedestrian' video sequence

T_c clock cycles) and the number of active APs in each iteration ($W_a[i]$) for both the Exe and Know heuristics in the first 100 iterations. Both the Exe and Know heuristics adapt to the variation in workload; however, the Know heuristic adapts better than the Exe heuristic. Firstly, the Know heuristic responds more quickly to sudden variations in workload due its proactive nature resulting from the use of the application's knowledge as marked in Fig. 6.7a. Secondly, the Know heuristic changes the number of active APs more often than the Exe heuristic to better keep up with the true workload. The throughput of the ME stage is reported in Fig. 6.7b for the first 100 iterations where the y-axis plots the clock cycles of each iteration including the overhead of the execution of the heuristics ($CC[i]$). It is obvious that the Exe heuristic incurred a significant penalty (up to 55,000 clock cycles) when the workload changed suddenly and significantly. On the other hand, the Know heuristic incurred a small penalty by activating a number of APs in advance due to the workload prediction from the application's knowledge. Thus, the Exe heuristic will result in more degradation of the throughput compared to the Know heuristic due to its reactive nature. Figures 6.8, 6.9, 6.10, and 6.11 show similar trends for the other video sequences as well.

Table 6.2 summarises the results of the comparison of the two heuristics. The second column reports the average number of active APs which is less than AP_M

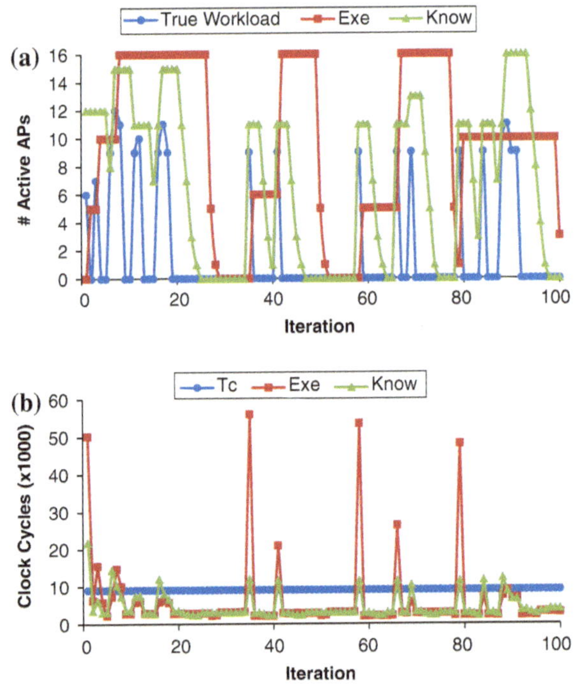

Fig. 6.8 **a** Adaptability and **b** Throughput for the 'sky' video sequence

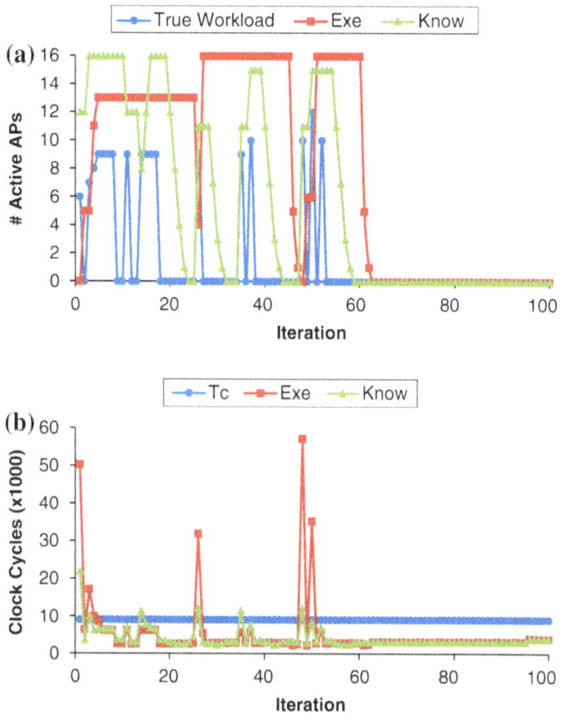

Fig. 6.9 **a** Adaptability and **b** Throughput for the 'station' video sequence

Table 6.2 Comparison of Exe and know heuristics

Video Sequence	Avg. active APs		T_c Violation (%)		Min. throughput (fps)		Avg. throughput (fps)	
	Exe	Know	Exe	Know	Exe	Know	Exe	Know
Pedestrian	13	10	3	11	24.17	28.76	25.42	29.00
Sky	10	8	5	8	23.22	28.76	23.98	29.10
Station	5	4	3	4	25.40	29.67	25.53	29.72
Sunflower	12	10	4	9	24.50	28.79	24.74	28.94
Tractor	6	5	4	5	24.93	29.53	25.12	29.56

(16 in the experiments) for all the video sequences, indicating the possibility of significant energy reduction. The third major column, termed 'T_c Violation', reports the number of iterations (as a percentage of the total iterations) that took more than T_c clock cycles, hence violating the throughput constraint. For example, 11 % of the iterations in the 'pedestrian' video sequence exceeded T_c clock cycles. The impact of such violations on the throughput of the adaptive pipelined MPSoC is reported in the fourth and fifth major columns. For example, the minimum and average throughput for 'pedestrian' video sequence is 28.76 and 29.00 fps respectively, which is the worst

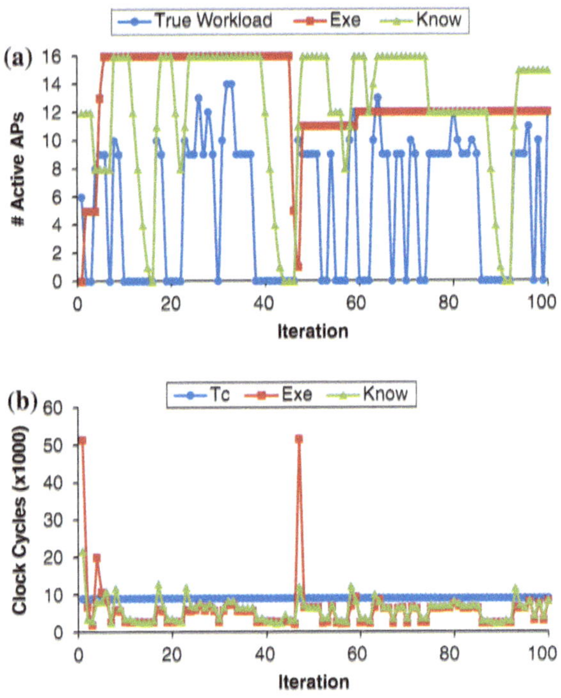

Fig. 6.10 **a** Adaptability and **b** Throughput for the 'sunflower' video sequence

amongst all the video sequences. It is interesting to note that the Exe heuristic uses more APs and incurs less number of throughput violations than the Know heuristic, yet it degrades the throughput more than the Know heuristic. The primary reason is that the Exe heuristic does not activate/deactivate APs at the right instants during the execution of the application due its reactive nature. Therefore, even with less number of throughput violations, each violation had a significant throughput penalty. To summarise, the Know heuristic incurs minimal degradation of the throughput due to a combination of run-time workload monitoring, workload prediction and off-line statistical analysis.

Let us now have a look at the energy reduction due to the proposed processor manager compared to a worst-case pipelined MPSoC where all the APs are always active. Figure 6.12 summarises the findings. The light and dark bars refer to the energy reduction using clock- and power-gating for deactivation of the idle APs respectively. The results show an energy reduction of up to 35 and 39 % with a minimum of 14 and 9 % for clock- and power-gating respectively, when the Know heuristic is used. These energy reductions were computed from the total energy consumption of the pipelined MPSoC (only the energy consumption of the processors was considered) rather than just the ME stage, *including the energy overhead of activation/deactivation of an idle AP and the run-time processor management heuristics*. Note that the Exe

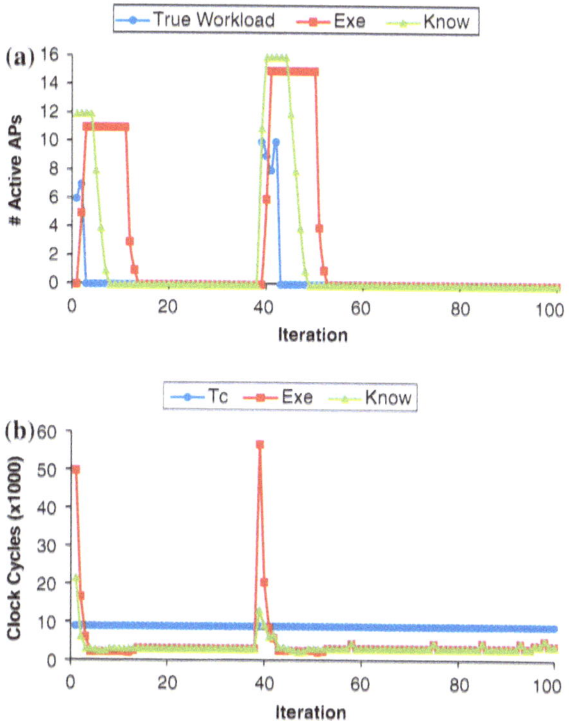

Fig. 6.11 **a** Adaptability and **b** Throughput for the 'tractor' video sequence

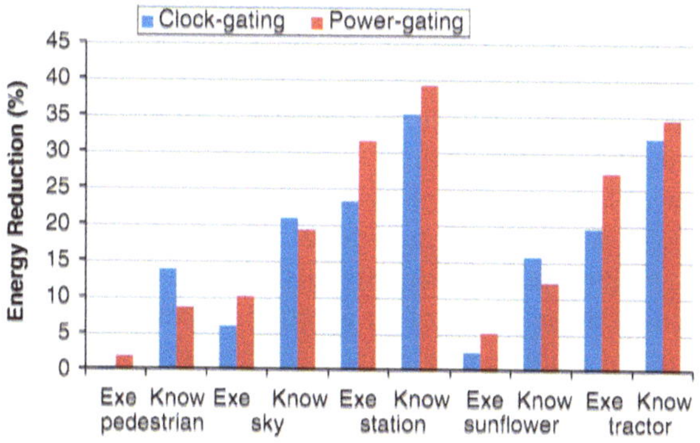

Fig. 6.12 Energy reduction of an adaptive pipelined MPSoC compared to a worst-case pipelined MPSoC

heuristic always saved less energy compared to the Know heuristic. In summary, the adaptive pipelined MPSoC with the Know heuristic delivered a minimum throughput of 28.75 fps with energy reduction of at least 9 % using either clock- or power-gating

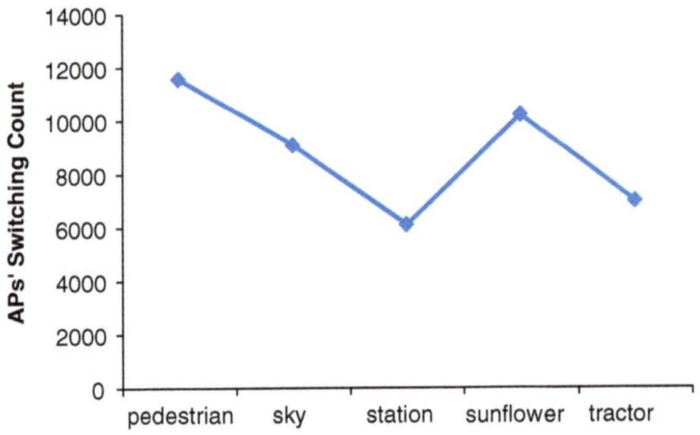

Fig. 6.13 Switching count of APs in the adaptive pipelined MPSoC

when compared to a worst-case pipelined MPSoC. Furthermore, the overhead of the processor manager is reasonable, illustrated by the fact that the throughput is not significantly degraded and the energy consumption of the adaptive pipelined MPSoC is reduced.

6.7.3 Discussion

Typically, power-gating results in more energy reduction than clock-gating as it reduces leakage energy in addition to dynamic energy. However, in Fig. 6.12, for the 'pedestrian', 'sky' and 'sunflower' video sequences, energy reduction from power-gating is lower than that of clock-gating. This is because power-gating incurs an overhead of activation/deactivation, which can increase significantly when the number of AP activations/deactivations increase. Figure 6.13 reports the total number of AP activations and deactivations (switching count) during the execution of the adaptive pipelined MPSoC. The switching count is significantly higher for the 'pedestrian', 'sky' and 'sunflower' sequences because these sequences exhibit high run-time workload variations, resulting in less energy reduction for power-gating. Therefore, the next chapter focuses on selection of an appropriate power reduction technique at run-time by the use of multiple power states and a power manager rather than blind use of either only clock-gating or only power-gating.

6.8 Summary

This chapter introduced an adaptive pipelined MPSoC architecture consisting of main processors and auxiliary processors for run-time adaptation to varying workloads. In addition, a distributed run-time processor manager was proposed to deactivate idle auxiliary processors, considering the variations in workload. By implementing an advanced multimedia application, the H.264 encoder supporting HD720p at 30 fps, the author illustrated that the adaptive pipelined MPSoC delivered a minimum throughput of 28.75 fps with energy reductions of up to 34 and 39 % for clock- and power-gating based deactivation of auxiliary processors respectively. These results show that adaptive pipelined MPSoCs provide an energy-efficient implementation platform for multimedia applications with run-time varying workload compared to worst-case pipelined MPSoCs.

References

1. H.264 test video sequences, http://media.xiph.org/video/derf/
2. International Telecommunucation Union, Advanced video coding for generic audiovisual services. Recommendation H.264 and ISO/IEC 14496—10:2005, 2005
3. Audio Video Coding Standard Workgroup of China, Audio video standard (avs), http://www.avs.org.cn/en/
4. H. Kalva, J.B. Lee, The vc-1 video coding standard. IEEE Multimedia **14**, 88–91 (2007)
5. S. Carta, A. Alimonda, A. Pisano, A. Acquaviva, L. Benini, A control theoretic approach to energy-efficient pipelined computation in mpsocs. ACM Trans. Embedded Comput. Syst. Article id 27: 6(4) 28 (2007)
6. A. Alimonda, S. Carta, A. Acquaviva, A. Pisano, L. Benini, A feedback-based approach to dvfs in data-flow applications. IEEE Trans. CAD Integr. Circ. Syst. **28**(11), 1691–1704 (2009)
7. H. Guo, S. Parameswaran, Balancing system level pipelines with stage voltage scaling, in *Proceedings of the IEEE Computer Society Annual Symposium on VLSI: New Frontiers in VLSI Design (ISVLSI '05)* (2005)
8. X. Liu, P.J. Shenoy, M.D. Corner, Chameleon: application-level power management. IEEE Trans. Mob. Comput. **7**(8), 995–1010 (2008)
9. W. Kim, M. Gupta, G.-Y. Wei, D. Brooks, System level analysis of fast, per-core dvfs using on-chip switching regulators, in *Proceedings of the IEEE 14th International Symposium on High Performance Computer Architecture (HPCA 2008)*, pp. 123–134 (2008)
10. K.K. Rangan, G. yeon Wei, D. Brooks, Thread motion: fine-grained power management for multi-core systems, in *Proceedings of the International Symposium on Computer, Architecture*, pp. 302–313 (2009)
11. M. Shafique, B. Molkenthin, J. Henkel, An hvs-based adaptive computational complexity reduction scheme for h.264/avc video encoder using prognostic early mode exclusion, in *Proceedings of the Design, Automation Test in Europe Conference Exhibition (DATE), 2010*, pp. 1713–1718, March 2010
12. M. Shafique, L. Bauer, J. Henkel, Enbudget: a run-time adaptive predictive energy-budgeting scheme for energy-aware motion estimation in h.264/mpeg-4 avc video encoder, in *Proceedings of the DATE*, pp. 1725–1730 (2010)
13. B. Zatt, M. Shafique, S. Bampi, J. Henkel, An adaptive early skip mode decision scheme for multiview video coding, in *Proceedings of the Picture Coding, Symposium* (2010)
14. Tensilica, Xtensa customizable processor, http://www.tensilica.com
15. T.-C. Chen, C.-J. Lian, L.-G. Chen, Hardware architecture design of an h.264/avc video codec, in *Proceedings of the 2006 Asia and South Pacific Design Automation Conference (ASP-DAC '06)* (IEEE Press, 2006)

16. M. Shafique, L. Bauer, J. Henkel, 3-tier dynamically adaptive power-aware motion estimator for h.264/avc video encoding, in *Proceedings of the ISLPED*, pp. 147–152 (2008)
17. J. Leverich, M. Monchiero, V. Talwar, P. Ranganathan, C. Kozyrakis, Power management of datacenter workloads using per-core power gating. Comput. Archit. Lett. **8**(2), 48–51 (2009)
18. T. Tuan, A. Rahman, S. Das, S. Trimberger, S. Kao, A 90-nm low-power fpga for battery-powered applications. IEEE Trans. Comput. Aided Des. Integr. Circ. Syst. **26**(2), 296–300 (2007)

Chapter 7
Power Management in Adaptive Pipelined MPSoCs

System-level power management schemes are often deployed in MPSoCs to exploit the idleness of processors at run-time for energy reduction by putting idle processors in low-power states [1, 2]. These schemes decide "when" and "which" power state should be selected for a processor to maximally reduce the energy consumption of the MPSoC. The decision is a challenging one due to the latency and energy overheads involved in a transition from one power state to another. The aim of this chapter is to propose a power manager for an adaptive pipelined MPSoC to select the most suitable power state for each of the idle auxiliary processors.

Most of the run-time power management schemes are categorised as predictive schemes and stochastic techniques [1]. Predictive techniques typically exploit temporal correlation between the past history of the workload and its near future to predict the upcoming workloads. On the other hand, stochastic techniques model the workload behaviour as a controlled Markov process, and then find the optimal power management scheme based on the model. Predictive techniques suffer when the workload varies suddenly and significantly [1], while stochastic approaches suffer from the inaccuracies in the workload model and the complexity involved in solving the optimisation problem at run-time [2]. These issues primarily limit the use of both the predictive and stochastic schemes to systems where either the workload is very regular or the workload model is known a priori. Some advanced history based heuristics and stochastic schemes have been shown to predict with high accuracy in varying workloads; however, their computational complexity severely limits their use [2] and may not be suitable for fine-grained run-time management (which is required by real-time multimedia applications to avoid degradation of the throughput). Hence, Liu et al. [3] proposed the use of application knowledge for efficient run-time power management schemes because the application by far knows (or may know) the most about its future workload. The experiments illustrated application-aware power management outperforming OS-level and hardware-level schemes. However, the work in [3] exploited only a limited application knowledge (algorithmic properties such as the size and type of the frames) in a uniprocessor system. In this chapter, the author leverages more diverse application knowledge (algorithmic and input data properties) for run-time power management in adaptive pipelined MPSoCs.

H. Javaid and S. Parameswaran, *Pipelined Multiprocessor System-on-Chip for Multimedia*, DOI: 10.1007/978-3-319-01113-4_7,

Table 7.1 Typical power states of a processor (CG and PG stand for clock-gated and power-gated respectively)

Power state	Description	Power consumption	Transition energy	Wake-up latency
0	Active	1	0	0
1	CG	0.4	0.01	0.01
2	Partially PG	0.1	0.4	0.6
3	Fully PG	0.01	1	1

The values of power consumption, transition energy and wake-up latency are normalised, and are inferred from [4]

7.1 Motivational Example

Table 7.1 shows four typical states available for a processor (where CG and PG stand for Clock-Gated and Power-Gated respectively). The values illustrate that the Power State 3 (PS3) will result in the most energy saving; however, the amount saved will depend on the amount of time the processor will remain in PS3, and this saving should amortise the energy overhead of the transition. Like Chap. 6, consider an adaptive pipelined MPSoC where the motion estimation stage is implemented with one *Main Processor (MP)* and sixteen *Auxiliary Processors (APs)* (designed for HD720p at 30 fps), where the APs can be deactivated using either only clock-gating or only power-gating. These sixteen APs are not active at all times, and are used only when the workload is beyond the capacity of the MP. Figure 7.1 shows the activity of one of the APs in the motion estimation stage, where 1 and 0 mean the AP is active and idle respectively. The figure shows that the idle periods (number of consecutive idle iterations) of the AP varies significantly at run-time. Power-gating will not be beneficial during short idle periods due to its relatively large wake-up overhead, while clock-gating will not be beneficial during long idle periods as it only saves dynamic power. Hence, both clock- and power-gating alone, as used in Chap. 6, do not exploit the full potential of idle periods because they do not evaluate the suitability of clock- and power-gating depending upon the duration of an idle period. On the other hand, a run-time power manager with the provision of multiple power states will provide a fine-grained power reduction knob as multiple power states [4]

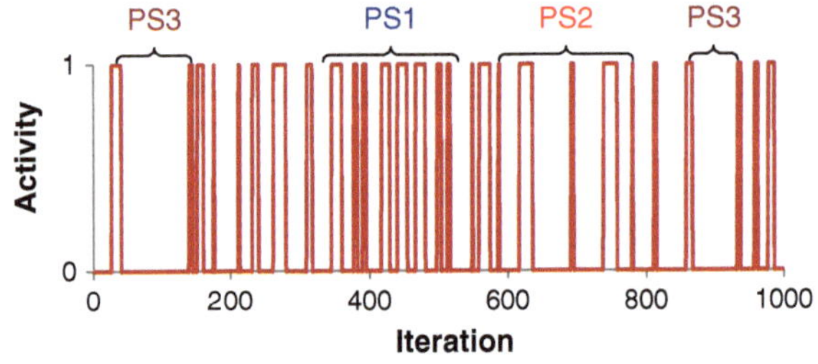

Fig. 7.1 Activity of one of the APs in the motion estimation stage of the H.264 video encoder

trade-off wake-up latency and energy with the possible energy savings in an MPSoC. For example, Fig. 7.1 illustrates that the AP is transitioned to different power states (PS1, PS2 and PS3 from Table 7.1) depending on the duration of the idle periods instead of always power-gating or clock-gating it, which will result in more reduction in energy consumption of an adaptive pipelined MPSoC. However, the challenge is to predict, with high accuracy and in the presence of run-time variations in workload, the duration of an upcoming idle period for an AP so that the most beneficial power state can be selected for it.

Therefore, this chapter builds upon the processor manager of the last chapter (which determined idle APs during every iteration) to propose a power manager for an adaptive pipelined MPSoC (to select the most suitable power state for each of the idle APs). Firstly, an analytical analysis is conducted so as to calculate the minimum number of iterations for a given power state to be energy-wise beneficial for an AP. That is, the given power state will be energy-wise beneficial if the AP stays in that power state for at least the minimum number of iterations of that power state. Secondly, five heuristics are proposed as part of the power manager to decide, at run-time, the most beneficial power state for an idle AP. These heuristics attempt to forecast the duration of an upcoming idle period of an AP using either the application's execution history or knowledge. Then, based on the predicted duration of the idle period, the most suitable power state is selected for an idle AP.

7.2 Power Manager

Figure 7.2 shows a typical adaptive pipelined MPSoC, which is comprised of various pipeline stages. The adaptable stages are implemented with a combination of MPs and APs while the non-adaptable stages are implemented with MPs only. This makes adaptive pipelined MPSoCs an effective platform for advanced multimedia applications which contain stages with both almost constant workload and run-time varying workload. The architecture of the adaptive pipelined MPSoC allows for both a centralised and a distributed power manager. Like the processor manager in Chap. 6, a distributed power manager is proposed where an MP monitors and controls its own APs, independent of other MPs. Therefore, the power manager can be tweaked for each stage of the adaptive pipelined MPSoC. For example, differing stages can have differing power states for the APs depending on the type of processors used, in addition to different power management heuristics. Note that stages with almost constant workload do not need any power manager; thus, avoiding run-time overheads for such stages.

Figure 7.2 zooms in on one of the MPs to illustrate the two run-time managers, which are used in an adaptive pipelined MPSoC. The first manager, named the processor manager and described in detail in Chap. 6, decides "when" and "how many" APs to activate and deactivate. The APs to be deactivated (that is, the idle APs) are determined at the start of each iteration and remain deactivated until the end of the current iteration (that is, for T_c clock cycles). The second manager, named the power

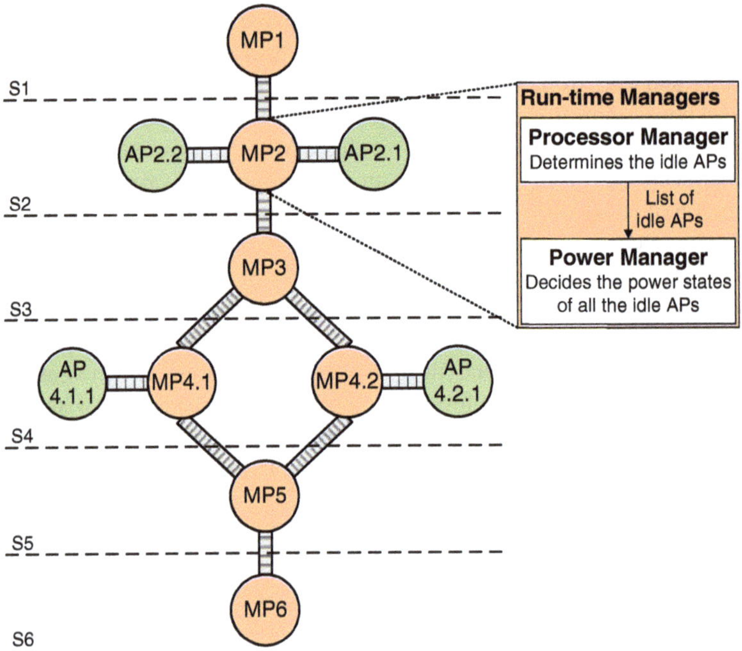

Fig. 7.2 Adaptive pipelined MPSoC's architecture with run-time managers

manager, then decides the power state of all the idle APs based upon the durations of the idle periods predicted for them. For example, if the processor manager reports AP14 and AP15 to be idle during the current iteration, then the power manager will decide the power states of AP14 and AP15 to maximally reduce the energy consumption of the adaptive pipelined MPSoC. This chapter uses the Know heuristic (Algorithm 4) from Chap. 6 in the processor manager. Note that an iteration of a processor refers to processing of one input data unit, where an iteration is considered idle if the processor is inactive during it.

7.2.1 Analytical Analysis

The decision on the power state of an idle AP depends on the duration of its idle period, that is, the number of consecutive idle iterations of the AP. In this section, the author shows an analytical method to calculate the *minimum number of iterations* for each power state so that if an AP is transitioned to a given power state for at least the minimum number of iterations, then the transition would be energy-wise beneficial. For the purpose of analysis, the following terms are introduced (some of the terms were introduced in Chap. 6 and are reproduced here for ease of readability):

- N power states denoted as $PS_0, \ldots, PS_{N-1}$ with power consumptions $P_0, \ldots P_{N-1}$ respectively, where $P_i > P_j$ for $i < j$. Hence, PS_0 would be the active state while PS_{N-1} would be the most power saving state.
- E_i^{ov}: Energy overhead of switching from PS_0 to PS_i and back to PS_0. It is assumed that state PS_i directly transitions to PS_0 without any transition to intermediate states, which is similar to the assumption used in [2].
- T_i^{ov}: Wake-up latency from PS_i to PS_0. An AP is activated and then the data is sent to overlap the activation time with the communication time as the FIFOs between the MP and the APs are always active. It is assumed that the communication time is greater than the wake-up latency which is typically the case with complex multimedia applications (see Sect. 7.5.1). Hence, T_i^{ov} will not be the deciding factor for the minimum number of idle iterations.
- T_c: The throughput constraint of the multimedia application. The duration of an iteration will be T_c clock cycles for all the stages, and hence for all the MPs and APs.

The reduction in energy consumption in a transition from PS_0 to PS_i should amortise the overhead of the transition. The energy consumption of an AP for I iterations in PS_i would be:

$$P_i \times T_c \times I + E_i^{ov}$$

The first factor computes the energy consumption of PS_i for I iterations, while the second factor is the overhead of the transition to PS_i from PS_0 and back to PS_0. To evaluate whether a transition to PS_i (higher power state) or a transition to PS_j (lower power state) would be beneficial, the energy consumption in PS_j including the overhead should be less than the energy consumption in PS_i:

$$P_j \times T_c \times I + E_j^{ov} < P_i \times T_c \times I + E_i^{ov}$$

$$I > \frac{E_j^{ov} - E_i^{ov}}{(P_i - P_j) \times T_c}$$

where $0 \leq i < j < N$. Hence, if the number of idle iterations is more than $\frac{E_j^{ov} - E_i^{ov}}{(P_i - P_j) \times T_c}$, then the transition to PS_j (lower power state) would be beneficial, otherwise the idle AP should be transitioned to PS_i (higher power state). Thus,

$$I_{j,i} = \left\lceil \frac{E_j^{ov} - E_i^{ov}}{(P_i - P_j) \times T_c} \right\rceil$$

is defined as the minimum number of iterations for PS_j to be beneficial than PS_i. Consequently, a power state can be compared with all the high power states to obtain the values of $I_{j,i}$. The minimum number of iterations for a power state PS_j would then be:

$$I_j^{min} = \max\{I_{j,i} : \forall\, i \; and \; 0 \leq i < j < N\}$$

Table 7.2 Minimum number of iterations required for the power states described in Table 7.1

Power state	I_j, i	$I_j^{\min}$
0	–	–
1	$I_{1,0} = 1$	1
2	$I_{2,0} = 1$, $I_{2,1} = 2$	2
3	$I_{3,0} = 2$, $I_{3,1} = 3$, $I_{3,2} = 7$	7

Let us go through the example power states of Table 7.1 to illustrate the calculation of I_j^{min}. The results are shown in Table 7.2. For each power state, the values of $I_{j,i}$ are computed assuming $T_c = 1$ sec. The value of $I_{1,0}$ signifies the fact that an AP should only be transitioned to PS1 from PS0 if the AP will be idle for at least one iteration. It should be noted that a transition to PS2 from PS0 for one iteration will also be beneficial ($I_{2,0} = 1$); however, for PS2 to be beneficial than PS1, the AP should be in PS2 for at least 2 iterations ($I_{2,1} = 2$). Thus, if an AP remains in PS_j for at least I_j^{min} number of iterations, it is ensured that the energy saving would be more than the transition to any of the higher power states.

The minimum number of iterations for each power state is computed off-line and then saved in the MP for use at run-time by its power management heuristic. The power management heuristic will predict the number of idle iterations (let us say I_{idle}^p) for an AP at the start of an iteration, which will then be used to obtain the most beneficial power state from the saved values of I_j^{min}. For example, for values of 1, 5 and 8 for I_{idle}^p, the AP will be transitioned to PS1, PS2 and PS3 respectively.

7.2.2 Leveraging Application Knowledge

Like Chap. 6, a pre-processing stage is employed to leverage application knowledge. The pre-processing stage analyses the variance and brightness of macroblocks of the incoming video frames to categorise them according to the motion contained in them. The category of the macroblock is then used to select its workload range, where workload ranges for different categories of macroblocks are computed off-line and are saved in a lookup table for use at run-time. The selected workload range is considered the predicted workload of the corresponding iteration of the motion estimation stage in the adaptive pipelined MPSoC. For example, the pre-processing stage may categorise macroblocks as either low- or high-motion macroblocks, and the workload ranges for these two categories may be [0, 5] and [6, 20] number of

APs respectively. Note that the workload prediction is fuzzy as it is in the form of a range. Also note that application knowledge can be exploited in other ways for other stages of the adaptive pipelined MPSoC, if required.

7.3 Problem Statement

Given the minimum number of iterations for each power state and the number of idle APs for the current iteration, the goal is to select the most beneficial power state for the idle APs so as to maximally reduce the energy consumption of the system with minimal degradation of the throughput.

The challenge is to accurately predict the duration of an idle period for an AP because an incorrect prediction of idle period's duration may result in either less energy reduction or even an increase in energy consumption. Consider the scenario where the predicted duration is longer than the actual duration of the idle period. Then, the idle AP may be transitioned to a lower power state (according to the predicted duration); however, it will be activated before the end of the predicted duration (due to shorter actual duration). At this instant, it is quite possible that the energy overhead in transition has not yet been amortised by the energy saving from the actual duration of the idle period, resulting in an increase rather than reduction in energy consumption. On the other hand, if the predicted duration of the idle period is shorter than the actual duration, then the maximum energy saving will not be exploited as the idle AP might be transitioned to a higher power state. Thus, a sophisticated run-time manager is required to decide the most beneficial power state for an idle AP. In addition, such a run-time manager should have low performance and energy overheads.

7.4 Power Management Heuristics

This section describes five heuristics for run-time management of the power states of idle APs. The first heuristic uses only the application's execution history. The other four heuristics leverage the workload prediction from the application's knowledge. Note that the processor manager always decides the number of idle APs at the start of an iteration. Thus, all the heuristics described below transition the APs to their corresponding power states at the start of the iteration. For the sake of simplicity, the heuristics are explained from the perspective of one MP; however they are equally applicable to other MPs of the adaptive pipelined MPSoC. The following terms are introduced to explain the heuristics (some of the terms are reproduced from Chap. 6 for ease of readability):

- $W_a[i]$: Actual workload of the ith iteration, equal to the number of APs that are active at the end of the ith iteration. For the current iteration, $W_a[i]$ holds the number of currently active APs.

- AP_M: The total number of APs for an MP, where APs are denoted as AP0, AP1, ... AP$M-1$.
- $idleAPs$: The list of APs that will be idle during the current iteration, which is provided by the run-time processor manager.
- I^p_{idle}: The predicted duration of idle period (number of idle iterations) for an AP.
- I^{min}_j: The minimum number of iterations for power state PS_j as explained in Sect. 7.2.1.

7.4.1 Application Execution Based Heuristic (Exe Heuristic)

The Exe heuristic monitors the workload of the previous iterations to keep a record of the average duration of an idle period (average number of idle iterations) of each AP which is later used to predict the duration of an idle period for an AP. The algorithm is shown in Algorithm 5. The algorithm keeps the total number of idle iterations (*totalIdleIterations* array) and the total number of idle periods (*idlePeriods* array)

Algorithm 5: Exe Heuristic

```
// Initialisation
for i = 0; i < AP_M; i++ do
    idlePeriods[i] = 0;
    totalIdleIterations[i] = 0;

// Called at the start of k-th iteration to decide the power
states
for i ∈ idleAPs do
    I^p_idle = ⌊totalIdleIterations[i] / idlePeriods[i]⌋;
    for j = 1; j < N; j++ do
        if I^p_idle < I^min_j then
            break;
    Transition i^th idle AP to power state PS_{j-1}

// Called at the end of k-th iteration to populate the
history information
if W_a[k] ≤ W_a[k − 1] then
    for i = W_a[k]; i < AP_M; i++ do
        totalIdleIterations[i]++;
else
    for i = W_a[k − 1]; i < W_a[k]; i++ do
        idlePeriods[i]++;
    for i = W_a[k]; i < AP_M; i++ do
        totalIdleIterations[i]++;
```

$W_a[k]$		16	13	14	13	13	16
Iteration		k-3	k-2	k-1	k	k+1	k+2
totalIdle Iterations	AP15	200	201	202	203	204	204
	AP14	150	151	152	153	154	154
	AP13	120	121	121	122	123	123
idle Periods	AP15	25	25	25	25	25	26
	AP14	28	28	28	28	28	29
	AP13	20	20	21	21	21	22

Fig. 7.3 An example illustrating the working of the Exe heuristic

seen till the current iteration (kth iteration in the Algorithm 5) for all the APs. The application's execution information is populated at the end of the current iteration (lines 10–17), while this information is used at the start to choose the power state for idle APs (lines 4–9). The predicted duration of idle period, I_{idle}^{p}, is the average number of idle iterations so far (line 5), and is used to select the most beneficial power state using I_j^{min} (lines 6–9).

The application's execution information is populated as follows. If the current number of active APs ($W_a[k]$) is less than the previous iteration's active APs ($W_a[k-1]$), then the total number of idle iterations for all the inactive APs (including the ones which were rendered idle in the current iteration) is incremented by one (lines 10–12). On the other hand, if $W_a[k-1] > W_a[k]$, then some of the APs were activated in the current iteration, and for these APs the number of idle periods (because the idle period of these APs has just finished) is incremented by one (lines 14 – 15). For the rest of the APs, the total number of idle iterations in incremented by one (lines 16–17). An example illustrating the working of the algorithm is shown in Figure 7.3, where $AP_M = 16$ and the calculation is shown for only the last three APs. At iteration $k-2$, consider $idleAPs = \{13, 14, 15\}$. Then, the Exe heuristic will put AP15 ($I_{idle}^{p} = \lfloor 200/25 \rfloor = 8$) to PS3 (since $8 \geq I_3^{min}$, see Table 7.2), while AP14 ($I_{idle}^{p} = \lfloor 151/28 \rfloor = 5$) and AP13 ($I_{idle}^{p} = \lfloor 121/20 \rfloor = 6$) will be transitioned to PS2.

It should be noted that the Exe heuristic keeps the minimum amount of information so that its run-time overhead is low. Furthermore, the average duration of idle period for each AP is updated at run-time based on the application's execution; however, the Exe heuristic will not be able to predict an idle period very accurately due to sudden variations in workload (see Sect. 7.5.2).

7.4.2 Application Knowledge Based Heuristics (Know Heuristics)

As explained in Sect. 7.2.2, a pre-processing stage is available which can predict the workload of an iteration beforehand in the number of APs that will be required during that iteration. In this section, the author shows how that prediction can be used to predict the duration of idle periods for APs, and then the author shows how

	$W_p[i]$	16	13	14	13	13	16
	Iteration	i	i+1	i+2	i+3	i+4	i+5
	AP15	5	4	3	2	1	
idle	AP14	5	4	3	2	1	
Periods	AP13	2	1	3	2	1	
	AP12	1	1	1	1	1	

Fig. 7.4 An example of populating idlePeriods table

the predicted duration of an idle period is used by the heuristics to decide the power states of idle APs. The following terms are used in addition to the ones described in Sect. 7.4 (some of the terms are reproduced from Chap. 6 for ease of readability):

- $W_p[i]$: Predicted workload for the ith iteration in number of required APs from the pre-processing stage.
- *idlePeriods[k][i]*: At ith iteration, the duration of the idle period (number of consecutive idle iterations) for the kth AP. For example, *idlePeriods[0][10] = 3* means that the duration of the idle period starting at iteration 10 for AP0 is 3 iterations. That is, starting at iteration 10, AP0 will remain idle for 3 iterations until iteration 12. This table is populated using the workload prediction from the pre-processing stage.
- MB_N: The total number of macroblocks in a frame. Although this information is specific to video encoder/decoder applications; however, it is used for ease of understanding and is not a limitation of the proposed heuristics. This information can be generalised as the maximum number of data units (iterations) in a multimedia application that can be pre-processed in advance for extraction of useful information and workload prediction for the run-time managers.

The example in Fig. 7.4 illustrates the computation of *idlePeriods* table for the last four APs only where $AP_M = 16$. The idea is to look into the future iterations to compute the duration of idle period of an AP, if it is deactivated at the start of the current iteration. For example, at iteration i in Fig. 7.4, if AP15 is deactivated, then it will be idle for the next 5 iterations according to the workload prediction because it will be activated again in iteration $i + 5$ (when $W_p[\text{i+5}] = 16$). Hence, the predicted duration of idle period for AP15 at iteration i will be 5. As another example, AP12 will be idle for only 1 iteration as it will be used in iteration $i + 1$ according to the predicted workload.

The algorithm to populate the entries of the *idlePeriods* table is shown in Algorithm 6. It populates the entries for the ith iteration (ith column of the table) based on the $(i - 1)$-th iteration's values and predicted workloads of future iterations. The initialisation is done at the first iteration (line 2) where the first column of the table is populated. Lines 4–10 look into the future iterations until the future workload equals the maximum number of APs (lines 9–10) to calculate the duration of idle period for all the APs. The variable pWH in lines 7–8 tracks the number of APs for which the duration of idle period has already been computed. For example, the first run of the for-loop in lines 5–6 will compute the duration of idle period for AP0–AP12 (since

Algorithm 6: Populate idlePeriods Table (for the sake of simplicity, boundary cases are not reported here)

```
1  for i = 0; i < MB_N; i++ do
2      if i == 0 then // First iteration
3          pHW = 0;
4          for ii=i+1; i < MB_N; ii++ do
5              for k = pHW; k < W_p[ii]; k++ do
6                  idlePeriods[k][i] = ii − i;
7              if W_p[ii] > pHW then
8                  pHW = W_p[ii];
9              if W_p[ii] == AP_M then
10                 break;
11     else
12         for k = W_p[i]; k < AP_M; ii++ do
13             idlePeriods[k][i] = idlePeriods[k][i-1] - 1;
14         pHW = 0;
15         for ii=i+1; i < MB_N; ii++ do
16             for k = pHW; k < W_p[ii]; k++ do
17                 idlePeriods[k][i] = ii − i;
18             if W_p[ii] > pHW then
19                 pHW = W_p[ii];
20             if W_p[ii] ≥ W_p[i] then
21                 break;
```

$i = 0$, $W_p[\mathrm{i}+1] = 13$). The second run of the same for-loop will only compute the idle iterations for the rest of the APs, that is, AP13 and onwards.

The second part of the algorithm (lines 11–21) populates the rest of the columns of the *idlePeriods* table. In this part, the duration of the idle period for some of the APs can be inferred from the previous iteration's values (lines 12–13). For other APs, the algorithm looks into the predicted workloads of future iterations until the future workload is the same or higher than the current iteration's workload (lines 20–21) to compute the duration of idle period (lines 14–15). For example, in Fig. 7.4, the values for AP14 and AP15 at $i + 2$ are computed by subtracting one from the values of $i + 1$ iteration; however, the values for AP12 and AP13 are computed from future workloads. Handling of the boundary cases and some optimisation steps are omitted for the sake of simplicity in Algorithm 6.

Once the *idlePeriods* table is available, the decision for the Know heuristic is simplified. Consider that the Know heuristic has to decide the power state for AP0 at the start of the kth iteration, then the value of *idlePeriods[0][k]* (which will be the predicted duration of the idle period for AP0) will be used to decide the most beneficial power state for AP0. Algorithmically, it is stated in Algorithm 7. As an example, in Fig. 7.4, if $idleAPs = \{14, 15\}$ at $i + 2$ iteration, then both AP14

Algorithm 7: Know Heuristic

```
// Called at the start of k-th iteration to decide the power
states
for i ∈ idleAPs do
    I^p_idle = idlePeriods[i][k];
    for j = 1; j < N; j++ do
        if I^p_idle < I^min_j then
            break;
    Transition i^th idle AP to power state PS_{j-1}
```

(idlePeriods[14][i+2] = 3) and AP15 (idlePeriods[15][i+2] = 3) will be transitioned to PS2 (see Table 7.2).

Recall from Sect. 7.2.2 that the workload prediction from the pre-processing stage is fuzzy and is represented as a range. However, the algorithm to compute the *idlePeriods* table assumes a single value for the predicted workload. Thus, four different mapping functions to obtain a single value from the predicted workload's range are used, resulting in four versions of the Know heuristic. Consider that Min(R), Max(R), Avg(R) functions return the minimum, maximum and average values of a range R respectively. Consider Q ranges are available from the pre-processing stage, which are numbered from 1 to Q where Max(R_Q) $\leq AP_M$. The following text uses the ranges of [0, 5] and [6, 20] in the number of APs for low-(L) and high-motion (H) macroblocks (from Sect. 7.2.2), and $AP_M = 20$ for exemplary purposes.

1. **MinKnow:** $Min(R_i)\ \forall\ i$ is used to map ranges to single values. For example, for a sequence of [L L H L] macroblocks, the predicted workloads would be [0 0 6 0]. The drawback of computing the *idlePeriods* table with Min(R) is that the maximum value of the predicted workload will be Min(R_M). This means that all the APs from Min(R_M) to AP_M-1 will always be considered inactive according to the predicted workloads. For example, AP6–AP19 will always be idle and hence will always be transitioned to PS3 (the most power saving state from Table 7.2).
2. **MaxKnow:** $Max(R_i)\ \forall\ i$ is used to map ranges to single values. For the same example of [L L H L] macroblocks, the predicted workloads would be [5 5 20 5]. Unlike MinKnow, MaxKnow introduces an error towards the other end of the spectrum. Since the minimum value of the workloads will be Max(R_1), the first Max(R_1) APs will be considered active during all the iterations according to the workload prediction. For example, AP0–AP4 will be active at all times and hence will only be transitioned to PS1 (the least power saving state from Table 7.2).
3. **AvgKnow:** $Avg(R_i)\ \forall\ i$ is used to map ranges to single values. For example, the predicted workload for the sequence of [L L H L] macroblocks would be [3 3 13 3]. In AvgKnow, all the APs from 0 to Avg(R_1)-1 (AP0 – AP2) will always be transitioned to PS1, while all the APs from Avg(R_M) to AP_M-1 (AP13 – AP19) will always be switched to PS3.

4. **FusedKnow:** $Min(R_1)$, $Avg(R_i)$, $Max(R_M)$ $\forall i, i \neq 1, i \neq M$ are used for the ranges. FusedKnow fuses the minimum of the first range, the maximum of the last range and the average of the intermediate ranges to compute the predicted workloads. For example, the sequence of [L L H L] macroblocks would be translated to [0 0 20 0]. FusedKnow will not suffer from the drawbacks of MinKnow, MaxKnow and AvgKnow as it uses $Min(R_1)$ and $Max(R_M)$ for the first and the last range respectively.

All these heuristics have to compute the *idlePeriods* table at run-time which might introduce an unacceptable overhead. The authors solution to this problem is to execute the table computation algorithm at the pre-processing stage. The pre-processing stage will write the table into a shared memory from which Know heuristic will read the values at run-time, keeping its overhead to a minimum. The computation of the *idlePeriods* table in the pre-processing stage will not affect the throughput of the video processing system as the pre-processing stage is not part of the multimedia system (see Fig. 7.5).

7.4.3 System-Level Overview

The system-level implementation of the proposed adaptive pipelined MPSoC with the run-time managers, executing a multimedia application such as H.264 video encoder, is shown in Fig. 7.5. Like Chap. 6, a multimedia application is implemented as a combination of pre-processing and multimedia systems. The pre-processing system extracts the features of incoming frames to provide useful information to the multimedia system for run-time adaptation. For example, the pre-processing stage can categorise the macroblocks according to the motion contained in them, as described in Sect. 7.2.2. The pre-processing stage is also responsible for the computation of the *idlePeriods* table using Algorithm 6. The multimedia system implements the video codec on an adaptive pipelined MPSoC. Each MP with a pool of APs implements run-time processor and run-time power managers. More specifically, the processor manager uses either the Exe heuristic or the Know heuristic from Chap. 6 to determine the idle APs at run-time. The power manager uses either the Exe heuristic or the Know heuristic described in Sects. 7.4.1 and 7.4.2 to decide the power states of idle APs at run-time. The pre-processing system is expected to work at the frame-level so that the predicted workload of all the macroblocks of a frame is available to the multimedia system which is working at the macroblock-level.

7.5 HD720p H.264 Video Encoder Case Study

In this section, the applicability of the proposed run-time power manager is illustrated by implementing an H.264 video encoder on an adaptive pipelined MPSoC supporting HD720p at 30 fps.

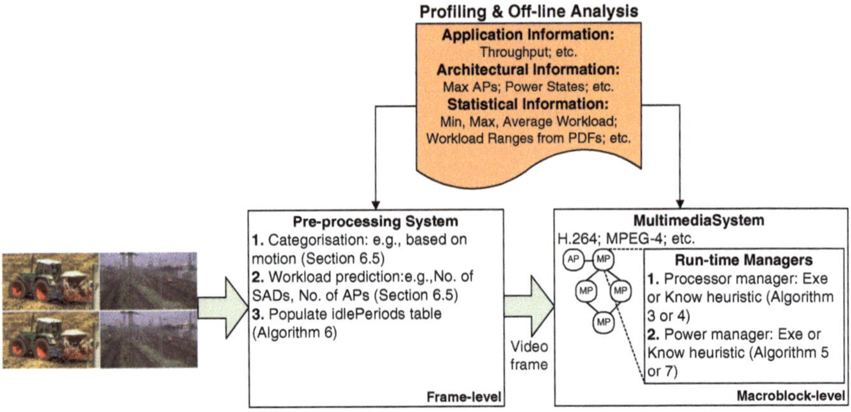

Fig. 7.5 A system-level implementation overview of run-time managers in adaptive pipelined MPSoCs

7.5.1 Implementation Details

The adaptive pipelined MPSoC created for the H.264 video encoder in Chap. 6 is used in this chapter as well. For proof of concept, both the processor and power managers were implemented for only the motion estimation stage. The motion estimation stage contained one MP and sixteen APs, running at a frequency of 1 GHz. Energy consumption of the adaptive pipelined MPSoC was measured by configuring the processors for a given 45nm technology.

The three power states shown in Table 7.3 were used for the APs. The values of transition energy and wake-up latency were inferred from [5, 6], while the values of I_j^{min} were computed according to the equations described in Sect. 7.2.1 with P_{dyn} = 28.5 mW, P_{leak} = 6.50 mW and T_c = 9,100 clock cycles (to support ≥30 fps). The adaptive pipelined MPSoC was tested with several video sequences to obtain average values of P_{dyn} and P_{leak} of an AP. In the adaptive pipelined MPSoC, the latency of sending the data (at least 256 ns assuming a byte transfer takes at least 1 clock cycle @ 1 GHz, see Sect. 6.7.1) to APs after activating them was larger than the wake-up latency of PS2 (100 ns) and hence did not affect the throughput of the pipelined MPSoC.

Table 7.3 Power states of the processors in the adaptive pipelined MPSoC

Power state	Power consumption	Transition energy (nJ)	Wake-up latency (ns)	I_j^{min}
0	$P_{dyn} + P_{leak}$	0	0	–
1	P_{leak}	1	3	1
2	~ 0	250	100	9

Like Chap. 6, the pre-processing stage categorised all the macroblocks of a frame into low-, medium- and high-motion macroblocks at run-time. The workload of each category in the number of APs was computed using offline analysis and was saved in a lookup table for use at run-time. The ranges (R_1, R_2 and R_3) for predicted workload of low-, medium- and high-motion macroblocks were [0, 4], [5, 10] and [11, 16] (number of APs) respectively. These ranges were also used by Algorithm 6 to compute the *idlePeriods* table for the Know heuristic.

7.5.2 Results and Analyses

The proposed power manager was tested with five different HD720p video sequences: pedestrian; sky; station; sunflower; and, tractor. Firstly, the capability of each heuristic in choosing the correct power state for the idle APs is illustrated. Figure 7.6 shows part of the whole results where the power state of AP0 and AP15 is plotted for the first 250 iterations for each of the five heuristics when the 'pedestrian' video was inputted to the adaptive pipelined MPSoC. Several notable facts are illustrated in the figure with labels A–E:

- **Label A** illustrates the scenario of incorrect power state transitions by the Exe heuristic. The duration of the idle periods pointed by the first two arrows is less than nine iterations. Hence, AP0 should have been transitioned to PS1; however, the average duration of an idle period according to the current state of the application's execution information was more than nine iterations. Thus, the Exe heuristic transitioned AP0 to PS2 which is not beneficial. The last arrow points out the converse scenario. Due to the recent short idle periods, the average duration of idle periods (from application's execution) dropped below nine, resulting in AP0's transition to PS1 instead of PS2 (the correct power state).
- **Label B** illustrates the drawback of the MaxKnow heuristic. Recall from Sect. 7.4.2 that the MaxKnow heuristic considers the first Max(R_1) APs (AP0–AP3 in the adaptive pipelined MPSoC) active during all the iterations. Thus, it is always switching AP0 to PS1 (the least power saving state) irrespective of the duration of the idle period.
- **Label C** illustrates the problem with the MinKnow heuristic. In the MinKnow heuristic, AP11–AP15 (Min(R_3) to AP_M-1) will be considered inactive at all times. Hence, AP15 is always transitioned to PS2 (the most power saving state) according to the MinKnow heuristic.
- **Label D** shows the scenarios where the AvgKnow heuristic will take wrong decisions on the power state of an AP. Since the AvgKnow heuristic uses Avg(R) for converting the ranges to single values, it will always consider AP0 as active and AP15 as inactive resulting in their transitions to PS1 and PS2 respectively irrespective of the idle periods' durations.
- **Label E** illustrates the scenario where the fuzzy workload prediction from the pre-processing stage can be misleading. AP15 should have been transitioned to

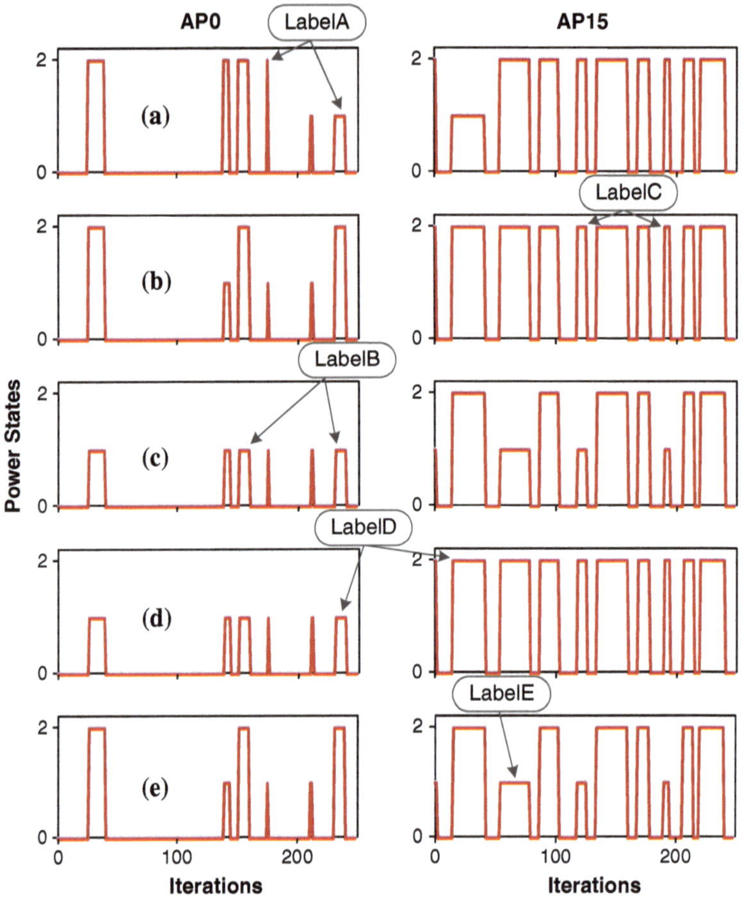

Fig. 7.6 Power states of AP0 and AP15 for the 'pedestrian' video sequence for (**a**) Exe, (**b**) MinKnow, (**c**) MaxKnow, (**d**) AvgKnow, and (**e**) FusedKnow heuristics

PS2 as the duration of the idle period is more than nine iterations, instead it was transitioned to PS1. Frequent wrong decisions on the appropriate power state might result in increased energy consumption; however, it is shown later that the number of wrong decisions from the FusedKnow heuristic is very low. This can also be seen from the graphs where the FusedKnow chose the wrong power state only once, that is, at Label E.

Figure 7.6 illustrates that FusedKnow performs the best in selecting the most beneficial power state for the two APs. Other APs in the adaptive pipelined MPSoC and other video sequences exhibited similar trends. Note that the use of only clock-gating and only power-gating in the processor manager of Chap. 6 would have always resulted in transition of both AP0 and AP15 to PS2 and PS1 respectively.

Table 7.4 Percentage error in the selection of power states by the power management heuristics when compared to the optimal scenario

Video sequence	Exe	Min know	Max know	Avg know	Fused know
Pedestrian	15.35	27	7.05	20.23	1.57
Sky	52.13	20.94	14.5	21.10	1.64
Station	47.28	18.03	25.52	21.64	2.53
Sunflower	26.41	24.51	18.09	20.75	1.43
Tractor	48.29	18.85	18.96	23.77	0.68

To compare the accuracy of the heuristics, an "Optimal" scenario was created by using the true workload of the motion estimation stage as the predicted workload, where the true workload was obtained from actual execution of the adaptive pipelined MPSoC. The power states selected in the optimal scenario would be the most beneficial states because the exact durations of idle periods are known from the actual execution. The results are reported in Table 7.4. The values report the number of wrong decisions taken by a heuristic as a percentage of the total decisions taken by it. For example, the Exe heuristic took 15.35 % wrong decisions in the selection of the power states for the 'pedestrian' video sequence. The second column shows that the error of the Exe heuristic is quite high which corroborates the fact that the application's execution based heuristics do not perform well in a widely varying workload. Column 6, on the other hand, reports the error of the FusedKnow heuristic which is always less than 3 %. This shows that appropriate leveraging of application knowledge can significantly improve the efficacy of the run-time power management heuristics. Another interesting fact is that the FusedKnow heuristic achieved such an accuracy using only fuzzy workload predictions (ranges of predicted workloads). Availability of better predictions (for example, 10 ranges instead of 3) would have further improved the accuracy of the FusedKnow heuristic.

Let us now examine the energy reduction achieved by the five power management heuristics. The proposed heuristics were compared to the optimal scenario and the use of only Clock-Gating (CG) and only Power-Gating (PG) in the processor manager of Chap. 6. The relative energy reduction was measured to show the improvement achieved by the utilisation of the proposed heuristics. The relative energy reduction of a heuristic j was computed as:

$$\frac{E_j^r - \min\{E_i^r : \forall\, i\}}{\min\{E_i^r : \forall\, i\}}$$

where E_j^r is the energy reduction of heuristic j over a worst-case pipelined MPSoC. A worst-case pipelined MPSoC does not adapt itself at run-time, and hence all the processors in it are active at all times. The value of E_j^r for a heuristic was computed by subtracting the energy consumption of the adaptive pipelined MPSoC (which included the energy consumption of all the processors in the adaptive pipelined MPSoC, excluding the memories) when heuristic j was used from the energy

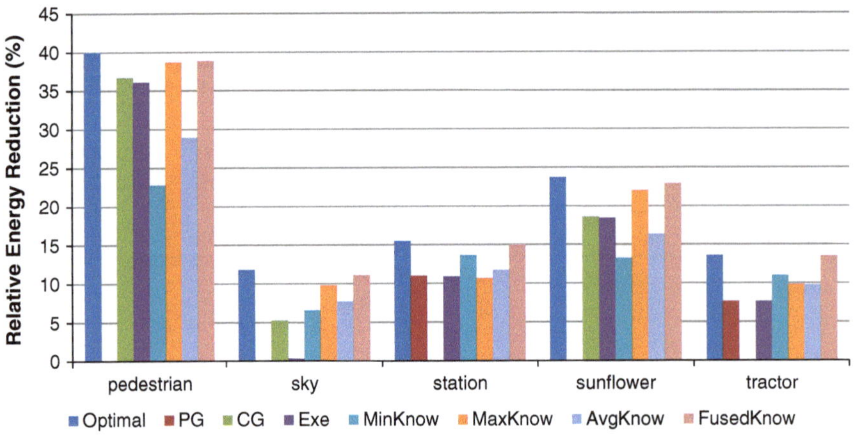

Fig. 7.7 Relative energy reduction of the power management heuristics in the power manager compared to the use of only Power-Gating (PG) or only Clock-Gating (CG) in the processor manager

consumption of the worst-case pipelined MPSoC. Therefore, the computed energy reduction *included the run-time energy overhead of the heuristics as well.* Interestingly, either only CG or only PG in the processor manager had the lowest energy saving for all the five video sequences, and thus the relative energy reduction depicted how much more energy was saved using the power manager compared to the use of the processor manager with naive power management.

Figure 7.7 reports the relative energy reduction achieved by the heuristics. For example, CG (the third bar) saved 36 % more energy than PG for the 'pedestrian' video sequence, while PG saved 11 % more energy than CG for the 'station' video sequence. It is obvious that the FusedKnow heuristic (the last bar) saves the most energy from amongst all the heuristics as it is closest to the optimal (the first bar) for all the video sequences. The FusedKnow heuristic was always within 1 % of the optimal result. This again shows the significance of proper leveraging of the application's knowledge at system-level for run-time power management. In terms of run-time performance overhead, it was found that the power manager degraded the throughput of the adaptive pipelined MPSoC by a maximum of 0.5 % compared to the use of only the processor manager. Hence, the effectiveness of the power manager can be seen from the fact that the FusedKnow heuristic saved up to 40 % ('pedestrian' sequence) more energy than the processor manager with only an additional throughput degradation of 0.5 %. This shows that adaptive pipelined MPSoCs with run-time processor and power managers provide a low-power implementation platform for multimedia applications.

7.6 Summary

In this chapter, a run-time power manager was proposed for adaptive pipelined MPSoCs. Five heuristics were proposed as part of the power manager to predict at run-time the upcoming idle period of an auxiliary processor and then to decide the most appropriate power state for it. These heuristics were guided by an analytical analysis and the application's execution or knowledge. A case study with an H.264 video encoder on an adaptive pipelined MPSoC showed that one of the application's knowledge based heuristics (FusedKnow) provided up to 40 % more energy saving with only a 0.5 % degradation of the throughput compared to a processor manager with naive power management (Chap. 6). These results show that the proposed run-time power manager is a feasible option in adaptive pipelined MPSoCs for low-power implementation of multimedia applications.

References

1. L. Benini, A. Bogliolo, G. De Micheli, A survey of design techniques for system-level dynamic power management. IEEE Trans. Very Large Scale Integration (VLSI) Syst. **8**, 299–316 (2000)
2. S. Irani, S. Shukla, R. Gupta, Online strategies for dynamic power management in systems with multiple power-saving states. ACM Trans. Embed. Comput. Syst. **2**, 325–346 (2003)
3. X. Liu, P.J. Shenoy, M.D. Corner, Chameleon: application-level power management. IEEE Trans. Mob. Comput. **7**(8), 995–1010 (2008)
4. K. Agarwal, K. Nowka, H. Deogun, D. Sylvester, Power gating with multiple sleep modes, in *Proceedings of the 7th International Symposium on Quality Electronic Design, ISQED '06*, pp. 633–637, 2006
5. J. Leverich, M. Monchiero, V. Talwar, P. Ranganathan, C. Kozyrakis, Power management of datacenter workloads using per-core power gating. Comput. Architect. Lett. **8**(2), 48–51 (2009)
6. T. Tuan, A. Rahman, S. Das, S. Trimberger, S. Kao, A 90-nm low-power fpga for battery-powered applications. IEEE Trans. Computer-Aided Des. Integr. Circuits Syst. **26**(2), 296–300 (2007)

Chapter 8
Multi-mode Pipelined MPSoCs

A pipelined MPSoC will typically be used as a multimedia accelerator because it is extremely customised for a specific multimedia application. Furthermore, it provides a programmable accelerator platform with reduced time-to-design and time-to-market because of the design automation methodologies described in Chap. 5 and in [1, 2].

Typical multimedia platforms (such as OMAP [3], Tegra [4], etc.) consist of a single-/multi-processor host system and multiple multimedia engines as shown in. The multimedia engines function as hardware accelerators, and are considered to be implemented as pipelined MPSoCs (rather than ASICs) in this chapter. The host system handles concurrent execution of general-purpose applications and off-loads multimedia applications to appropriate accelerators. It is often the case that not all the accelerators will be used simultaneously. For example, a user can either browse pictures or watch video in a smart phone, hence there is no need for concurrent execution of JPEG and H.264 decoders. Furthermore, various other video decoders such as MPEG-4 and VC-1 will not be required at the same time as H.264. Therefore, due to area constraints in portable media devices, it is desirable to use a multi-mode accelerator rather than individual accelerators when their use is mutually exclusive. If multiple applications need to be executed simultaneously such as listening to music while browsing pictures, then JPEG and MP3 decoders have to be implemented as two distinct accelerators. For example, Tegra [4] has five distinct accelerators Fig. 8.1.

The aim of this chapter is to reduce area footprint of pipelined MPSoCs based accelerators by combining mutually exclusive accelerators into a multi-mode accelerator with performance and energy consumption comparable to its individual counterparts. Therefore, multi-mode pipelined MPSoCs for multiple, mutually exclusive applications are proposed to function as multi-mode multimedia accelerators where each mode refers to the execution of one application. The author exploits the idea of merging individual application graphs (representing worst-case and adaptive pipelined MPSoCs) into a single application graph for realisation of a multi-mode pipelined MPSoC.

H. Javaid and S. Parameswaran, *Pipelined Multiprocessor System-on-Chip for Multimedia*, DOI: 10.1007/978-3-319-01113-4_8,

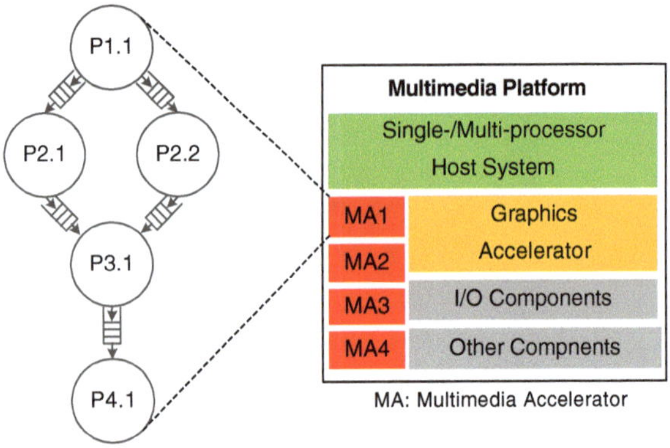

Fig. 8.1 A typical multimedia platform where multimedia accelerators are implemented as pipelined MPSoCs

Previous research has focused on resource sharing through merging of data-paths to reduce cost, area and power consumption of digital circuits. Initially, a number of works [5–7] formulated the problem of data-path merging as finding the maximum weight matching of a bipartite graph. However, bipartite matching based approaches mainly consider nodes in a data-path and ignore the edges. Moreano et al. [8] and Chong et al. [9] improved upon bipartite matching by formulating data-path merging problem in reconfigurable architectures and custom floating-point units as finding the maximum weight clique of a compatibility graph. Finally, Brisk et al. [10] formulated data-path merging problem in custom instructions as a substring/subsequence matching problem. However, their approach can only be applied to acyclic data-paths. In this chapter, the author builds upon the maximum weight clique approach to merge application graphs for a multi-mode pipelined MPSoC because: (1) both the nodes and edges of application graphs need to be merged to maximally reduce the number of processors and FIFO buffers and buffer sizes, and (2) multimedia application graphs can have cyclic dependencies.

Typical design of multi-mode systems [11–15] is done in two steps: first, selection of processing elements (from a given library) for application tasks; and second, mapping and scheduling of the tasks on the selected processing elements to minimize area, power, energy, etc. while ensuring designer constraints such as task deadlines, reliability constraint, etc. The works in [11–15] considered a predefined MPSoC platform with fixed number and types of processing elements. Hence, their problem was to select appropriate types of processing elements and then schedule the tasks on the selected elements to meet given task deadlines. On the other hand, in a multi-mode pipelined MPSoC, the number of processors is variable and depends on the application graphs. Thus, the problem here is to merge application graphs to maximally share processors and reduce area footprint. Like the problem of selection of processing elements, and mapping and scheduling of tasks, merging of application graphs is an NP-complete problem [16].

Merging of application graphs has been studied in [17–20]. The works in [17–19] mapped multiple applications, represented as Synchronous Dataflow Graphs (SDFs), on a tiled MPSoC architecture. They used a heuristic to merge multiple uses-cases of applications to reduce the number of tiles and number of links between the tiles. One of the heuristics proposed in this chapter (MaxN, see Sect. 8.4.2) is similar to their heuristic; however, they did not consider the size of buffers and different permutations of merging application graphs. Furthermore, two other heuristics are proposed where one of them finds optimal merging. Wildermann et al. [20] studied the mapping of multiple streaming applications on a tiled reconfigurable architecture. They merged application graphs using the technique in [10], and thus their work is limited to acyclic applications. Furthermore, their objective was to minimise FPGA reconfiguration time rather than the area footprint. Hence, unlike [17–20], three heuristics are proposed in this chapter to merge cyclic application graphs, trading-off their running time with optimality of the merged graph (in the context of pipelined MPSoCs). This is the first attempt at merging application graphs for multi-mode pipelined MPSoCs.

8.1 Multi-mode Pipelined MPSoCs

A multi-mode pipelined MPSoC is defined to support multiple, mutually exclusive applications by allowing several modes where each mode refers to the execution of only one of the applications on it. Two approaches can be taken to design multi-mode pipelined MPSoCs:

1. Individual pipelined MPSoCs designed separately for each application are merged at hardware-/gate-level through sharing of similar processors/FIFOs. Note that gate-level hardware sharing is often the norm for multi-mode ASICs, but might be infeasible for multi-mode pipelined MPSoCs due to high design complexity resulting from millions of gates in such pipelined MPSoCs.
2. An abstract, system-level representation of a pipelined MPSoC's architecture (number of processors, and number and connection of FIFO buffers) is described in a directed graph. Then, directed graphs representing individual pipelined MPSoCs are merged into a single graph by finding the maximal overlap between them. The merging of these graphs reveals system-level sharing of processors and FIFO buffers in the multi-mode pipelined MPSoC. This approach is further explored in this chapter.

The author uses application graphs to capture abstract, system-level representation of the pipelined MPSoCs. Since sub-kernels and edges in an application graph are one-to-one mapped to processors and FIFO buffers in a pipelined MPSoC respectively, an application graph inherently captures system-level representation of the pipelined MPSoC as a directed graph. Note that if a sub-kernel and edge of an application is mapped to multiple processors and FIFO buffers in the pipelined MPSoC, then both the sub-kernel and edge are replicated in the application graph accordingly to keep a one-to-one mapping between the application graph and the pipelined MPSoC. Thus, in the rest of the chapter, an application graph represents the abstract, system-level architecture of the pipelined MPSoC on which it will be executed. Note

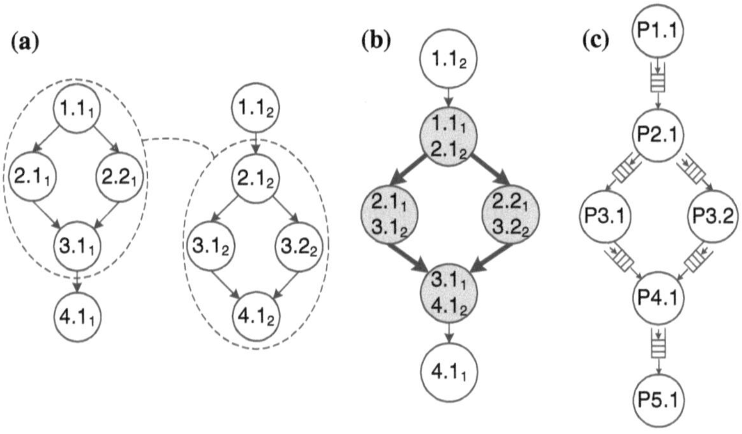

Fig. 8.2 Merging two application graphs to derive a multi-mode pipelined MPSoC: **a** Two application graphs **b** Merged application graph **c** Multi-mode pipelined MPSoC

that the application graphs abstract the types of processors (main processors, auxiliary processors, etc.), thus they can be used to merge both worst-case and adaptive pipelined MPSoCs.

Figure 8.2 shows an example of how two application graphs, representatives of individual pipelined MPSoCs, can be merged to derive a multi-mode pipelined MPSoC. The notation $m.n_x$ inside each node represents the nth sub-kernel in the mth stage of the xth application. For example, 3.1_2 represents the 1st sub-kernel in the third stage of the second application. The dotted lines illustrate one of the possible overlaps between the two application graphs. Based on the marked overlap, the combined application graph is shown in Fig. 8.2b where the grey coloured nodes represent the combined nodes of individual application graphs. The thick arrows show the merged edges from the individual application graphs. If each node and edge in the merged application graph is assigned to a processor and a FIFO buffer respectively, then a multi-mode pipelined MPSoC can be realised as shown in Fig. 8.2c. In mode 1, the first application will be executed using processors P2.1 ... P5.1 with processor P1.1 being idle. Likewise, in mode 2, only processors P1.1 ... P4.1 will be used to execute the second application. The multi-mode pipelined MPSoC uses only six processors and six FIFO buffers compared to a total of ten processors and ten FIFO buffers in individual pipelined MPSoCs.

The example above shows that multi-mode pipelined MPSoCs can reduce area footprint significantly; however, there are several issues that need to be addressed in driving a multi-mode pipelined MPSoC from the merged graph:

- The processors in individual pipelined MPSoCs contain custom instructions and cache configurations according to the sub-kernels mapped on them; however, the customisation information has been abstracted in the application graphs. Therefore, if two sub-kernels with differing custom instructions and cache configurations are merged, then the processor executing those sub-kernels in the multi-mode pipelined MPSoC will contain a union of all the custom instructions and the cache configurations. For example, the processor P2.1 will have the custom instructions for both 1.1_1 and 2.1_2 sub-kernels and the larger of the two cache configurations.

- For processors executing sub-kernels from multiple applications, the individual code segments of those sub-kernels are merged using a switch statement to select between the appropriate sub-kernel through the mode.
- In each mode, processors that do not belong to the currently executing application are power-gated to avoid an unnecessary increase in energy consumption of the application compared to its individual pipelined MPSoC counterpart.

The aim of this chapter is to design a multi-mode pipelined MPSoC (as defined above) with a minimal number of processors and FIFO buffers (due to the cost of wires and interconnects in ports) and buffer sizes for a set of applications by finding the maximum overlap among the application graphs. To this end, the following are assumed:

- Homogeneous multi-mode pipelined MPSoC for the purpose of merging application graphs. That is, sub-kernels of all the applications are executed on the same base processor. Heterogeneity is added after the merging process, where customisation from the individual pipelined MPSoCs is added to the multi-mode pipelined MPSoC (as explained above).
- The computation and communication ratios of the sub-kernels within each application or across different applications will not affect the balancing of the multi-mode pipelined MPSoC's modes. This is because: (1) a multi-mode pipelined MPSoC executes one application in a mode, and (2) the addition of custom instructions and cache configuration to its processors after the merging process balances its stages for each of its modes (because the individual pipelined MPSoCs were balanced through the same customisation of the processors). Therefore, computation and activation ratios do not need to be considered during the merging process.

8.2 A Design Flow

The design flow for creating a multi-mode pipelined MPSoC is as follows. The input consists of application graphs and their individual pipelined MPSoC implementations (obtained using the methods in Chap. 5 and/or Chap. 7). Firstly, application graphs are merged into a single application graph using one of the three merging heuristics proposed in Sect. 8.4. Secondly, the multi-mode pipelined MPSoC is derived from the merged application graph through one-to-one mapping. That is, each sub-kernel and edge is mapped to a processor and a FIFO buffer respectively. In the third step, the homogeneous multi-mode pipelined MPSoC (derived from merged graph) is balanced for each of its modes by utilising the customisation of the processors from individual pipelined MPSoCs. That is, if two sub-kernels with differing custom instructions and cache configurations are merged, then the processor executing those sub-kernels in the multi-mode pipelined MPSoC will contain the union of all the custom instructions and the cache configurations from the corresponding processors in the individual pipelined MPSoCs. Since the third step does not affect the merging of the application graphs, it is not discussed further in this chapter.

Note that multi-mode pipelined MPSoCs are used as accelerators and the applications are known apriori, thus they are optimised at design-time by merging application graphs and customisation of the processors. Hence, the overhead of run-time task mapping and merging techniques (which are used when the application mix is unknown at design-time) is avoided. At run-time, the host system will configure the multi-mode pipelined MPSoC in one of its modes to execute the desired application.

8.3 Problem Statement

An application is represented as a directed graph, G_x:

$$G_x = (V_x, E_x) : 1 \leq x \leq X$$

where X is the total number of applications. Each node in the set V_x is a sub-kernel, denoted as:

$$V_x = \{m.n_x : 1 \leq m \leq M_x, 1 \leq n \leq N_{m,x}\}$$

where M_x is the number of stages in the xth application graph and $N_{x,m}$ is the number of sub-kernels in the mth stage of the xth application graph. Each edge in an application graph denotes the data dependency between the sub-kernels and the amount of data that will be transferred in one iteration:

$$E_x = \{(m.n : i.j_x) : 1 \leq m, i \leq M_x, 1 \leq n \leq N_{m,x} \\ 1 \leq j \leq N_{i,x}\}$$

For example, the edge between 2.1_1 and 3.1_1 in Fig. 8.2 will be denoted as $2.1 : 3.1_1$. Each vertex $v_x \in V_x$ has a hardware implementation cost denoted as $P(v_x)$. Each edge $e_x \in E_x$ is implemented as a FIFO buffer. The size of the buffer depends on the capacity of the edge, denoted as $C(e_x)$, which is the amount of data transferred in one iteration and is known a priori because multimedia applications send and receive the same amount of data in each iteration. Hence, each edge $e_x \in E_x$ has a hardware implementation cost $F(e_x)$ which depends on the size of the buffer and the cost of the two ports used to connect it to the reading/writing processors.

The area of a pipelined MPSoC is the summation of the area of all the processors and FIFO buffers. Since an application graph is one-to-one mapped to derive a multi-mode pipelined MPSoC, its area is calculated as:

Given X application graphs, the goal is to merge them into one application graph, G_{MG}, such that the area of the multi-mode pipelined MPSoC derived from G_{MG}

is minimal. This is equivalent to maximally reducing the number of nodes (cost of processors) and the number of edges (cost of processor/FIFO ports) and their capacities (size of FIFO buffers) in G_{MG}.

8.4 Merging Heuristics

In this section, three methods are described to solve the problem of merging X application graphs. Two of these methods are based on greedy heuristics, *MaxS* and *MaxN*, while the third one, *MaxC*, is based on maximum weight clique based approach to find the optimal merging of the application graphs. Figure 8.3 shows the working of the three heuristics on two application graphs, G_1 and G_2, and will be used as an example in the rest of this section.

8.4.1 MaxS (Maximum Stages)

The MaxS heuristic, described in Algorithm 8, works on the principle of keeping the applications' topologies. It selects the maximum number of stages from all the graphs as the stages of the merged graph, G_{MG} (line 2). Then, within each of those stages, it selects the maximum number of nodes from the corresponding stages of all the graphs (line 4). Each node in the mth stage of G_{MG} is obtained by combining the corresponding nodes from the mth stage of all the graphs. For example, in Fig. 8.3c, two nodes are added to G_{MG} in the second and third stages because both the second stage of G_1 and the third stage of G_2 contain two nodes each. The first node in the second stage of G_{MG} ($2.1_1/2.1_2$) is a combination of the first nodes from the second stage of both G_1 and G_2 while the second node only contains 2.2_1 since there is only one node in the second stage of G_2.

The second part of MaxS adds appropriate edges to G_{MG} (lines 6–12). For each edge e_x in a graph G_x, a corresponding edge is added to G_{MG} if it does not exist in G_{MG}. If the corresponding edge already exists in G_{MG}, denoted as e_{MG}, then e_x is combined with e_{MG}. When two edges are merged, the higher of the two capacities is used (lines 11–12). For example, in Fig. 8.3c, the thick arrow marked $1.1{:}2.1_1/1.1{:}2.1_2$ represents a merged edge of G_1 and G_2 (the capacities are omitted for the sake of simplicity). The heuristic does not use the lesser of the two capacities for the merged edge because the throughput of one of the applications will be degraded significantly (see Chap. 4).

Figure 8.3c shows the merged graph from MaxS where the grey coloured nodes and thick arrows represent the overlap between the two application graphs. Thick arrows along with solid and broken arrows highlight the topology of G_1 and G_2 within G_{MG} respectively. The MaxS is a stark greedy heuristic yet it has reduced the total number of nodes and edges from nine and nine in G_1 and G_2 to six and seven

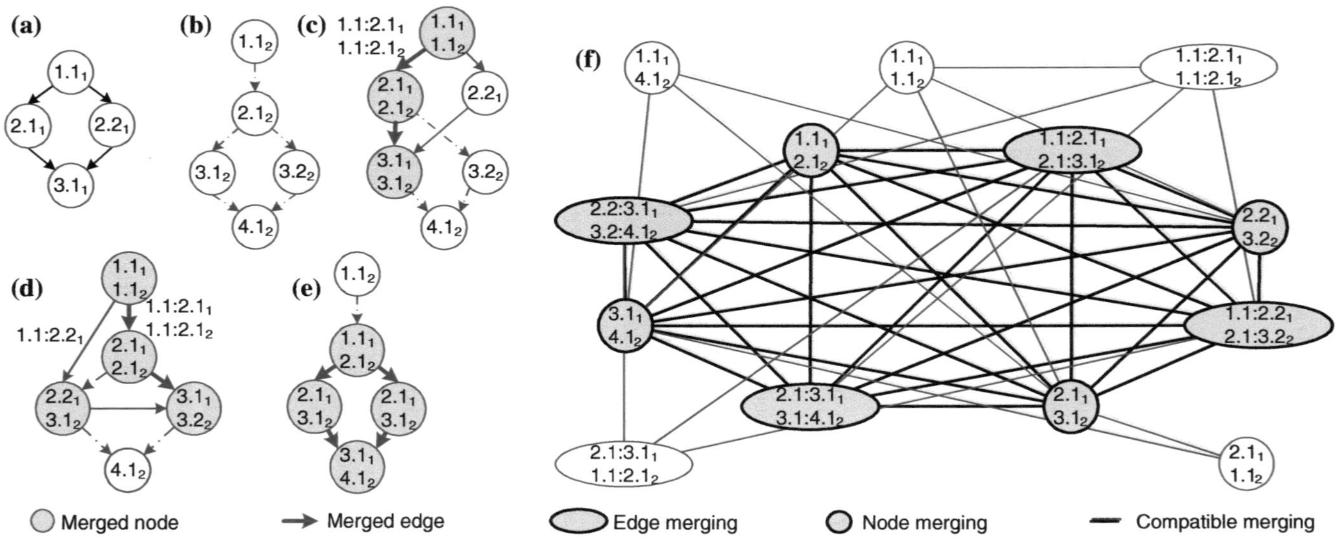

Fig. 8.3 Merging two application graphs: **a** G_1, **b** G_2, **c** G_{MG} from MaxS, **d** G_{MG}, from MaxN **e** G_{MG} from MaxC, **f** Compatibility graph, G_c, and maximum weight clique solution. For the sake of simplicity, edge capacities in G_1, G_2 and G_{MG}, and node weights in G_c are omitted

Algorithm 8: MaxS Heuristic

$G_{MG} = \emptyset$;
maxStages = $\max\{M_x\ :\ 1 \le x \le X\}$;
// Adding nodes to G_{MG}
for *m=1; m ≤ maxStages; m++* **do**
    $N_{m,MG} = \max\{N_{m,x} : 1 \le x \le X\}$;
    A node in mth stage of G_{MG} is combination of corresponding nodes from the mth stage of all G_x

// Adding edges to G_{MG}
for *x=1; x ≤ X; x++* **do**
    forall the $e_x \in E_x$ **do**
        if e_x *does not exit in* E_{MG} **then**
            Add e_x to E_{MG}
        else
            if $C(e_{MG}) < C(e_x)$ **then**
                $C(e_{MG}) = C(e_x)$;

respectively in G_{MG}. The MaxS heuristic visits all the nodes and edges in all G_x only once, and hence its complexity is $O(\sum_x |V_x| + |E_x|)$.

8.4.2 MaxN (Maximum Nodes)

Unlike MaxS, the MaxN heuristic focuses on maximally reducing the number of nodes in the merged graph. The algorithm is described in Algorithm 9 where the fundamental operation is to merge two graphs at a time (lines 4–12). Line 4 initialises G_{MG} with the graph that has the maximum number of nodes amongst G_{MG} and G_x. The reason is that the number of nodes in G_{MG} should not be greater than the maximum number of nodes from all the graphs.

Once G_{MG} is initialised, the algorithm traverses all the nodes in G_x in a breadth-first manner and combines its nodes with those of G_{MG} in a breadth-first manner as well (lines 5–6). For example, in Fig. 8.3d, 1.1_1, 2.1_1 and 2.2_1 are combined with 1.1_2, 2.1_2 and 3.1_2 respectively. After merging nodes, appropriate edges are added or merged by traversing all the edges in G_x (lines 7–12). Like MaxS, while merging edges, the higher of the two capacities is used (lines 9–10). For example, in Fig. 8.3d, the edge marked $1.1{:}2.1_1/1.1{:}2.1_2$ is combined from G_1 and G_2 while edge $1.1{:}2.2_1$ is added from G_1. The topology of G_1 and G_2 is illustrated with thick and solid arrows, and thick and broken arrows respectively. The MaxN heuristic has reduced the total number of nodes and edges from nine and nine in G_1 and G_2 to five and seven respectively in G_{MG}.

After merging the first two graphs, further graphs are combined with the already merged graph one by one (line 3). The amount of saving from MaxN depends on the order of merging the graphs. Hence, all the permutations of merging all the graphs are exhausted to select the merged graph with the minimum area (lines 1 and 13). For

Algorithm 9: MaxN Heuristic

forall the *permutations of merging all* G_x **do**
  $G_{MG} = G_1$;
  for x=2; $x \leq X$; x++ **do**
    if $\sum_m N_{m,x} > \sum_m N_{m,MG}$ **then**
      Swap G_{MG} with G_x
    `// Combining nodes and adding edges to` G_{MG}
    while *traversing* $v_x \in V_x$ *in breadth-first manner* **do**
      Combine v_x with $v_{MG} \in V_{MG}$ in breadth-first manner
    forall the $e_x \in E_x$ **do**
      if e_x *does not exit in* E_{MG} **then**
        Add e_x to E_{MG}
      else
        if $C(e_{MG}) < C(e_x)$ **then**
          $C(e_{MG}) = C(e_x)$;
return G_{MG} with minimum area

one merging of all G_x, MaxN visits the nodes and edges only once. Since there are $X!$ permutations of merging all G_x, the complexity of MaxN is $O(X! \sum_x |V_x| + |E_x|)$. This is reasonable as the number of applications is generally small (that is, $X < 10$) and the merging will be used only once at design-time.

8.4.3 MaxC (Maximum Weight Clique)

Unlike MaxS and MaxN, MaxC targets maximal reduction of both the nodes and edges in the merged graph. Reduction of the edges is important because addition of an edge costs a FIFO buffer and two ports (one for the writing processor and the other for the reading processor) which is expensive due to the extra memory required and associated area overhead of wires and interconnects.

The MaxC heuristic, shown in Algorithm 10, formulates the merging problem as a maximum weight clique problem. It consists of three parts: firstly, creating the compatibility graph, G_c (line 3); secondly, finding the maximum weight clique (line 4); and finally, constructing the G_{MG} (line 5). These operations are performed on the first two graphs and then subsequent graphs are merged with G_{MG} one by one (line 2).

The compatibility graph, $G_c = (V_c, E_c)$, is an undirected weighted graph that represents which node and edge merging of two graphs are compatible with each other. Each vertex $v_c \in V_c$ denotes either the merging of two nodes or two edges from G_x and G_y, and hence is annotated as (v_x/v_y) or (e_x/e_y). Each $v_c \in V_c$ has a weight w_c that corresponds to the area reduction achieved by that merging. The weights of all the vertices (v_x/v_y) will be $P(v_x)$ since merging two nodes would save

Algorithm 10: MaxC Heuristic

$G_{MG} = G_1$;
for x=2; $x \leq X$; x++ **do**
  Build G_c for G_{MG} and G_x
  Find maximum weight clique of G_c
  Reconstruct G_{MG}

a node in the merged graph and each node has the same weight $(P(v_x) = P(v_y))$. For (e_x/e_y) vertices, the weights are calculated as $\min\{C(e_x), C(e_y)\}$ because the edge with the higher capacity is used in the merged graph. An edge $e_c = (u_c, v_c) \in E_c$ indicates that the two merging represented by vertices u_c and v_c are compatible with each other. The edges are added according to the following rules:

- Vertices (v_x/v_y) and $(\dot{v}_x/\dot{v}_y)$ are compatible if $v_x \neq \dot{v}_x$ and $v_y \neq \dot{v}_y$. This means that a node in G_x cannot be merged with two different nodes in G_y.
- Vertices (e_x/e_y) and $(\dot{e}_x/\dot{e}_y)$ are compatible if $e_x \neq \dot{e}_x$ and $e_y \neq \dot{e}_y$. This means that an edge in G_x cannot be merged with two different edges in G_y.
- Vertices (v_x/v_y) and (e_x/e_y) are compatible if any of the following holds:
  - $src(e_x) == v_x$ && $src(e_y) == v_y$
  - $dst(e_x) == v_x$ && $dst(e_y) == v_y$
  - $src(e_x)$!= v_x && $src(e_y)$!= v_y
  - $dst(e_x)$!= v_x && $dst(e_y)$!= v_y

  where $src(e_x)$ and $dst(e_x)$ returns the source and destination node of e_x respectively. This rule means that the edges in G_x and G_y are merged only if their source and destination nodes are merged as well.

The compatibility graph for G_1 and G_2 is illustrated in Fig. 8.3f where circles and ovals represent possible node and edge merging respectively. For the sake of simplicity, only interesting vertices are shown.

To find the maximum overlap between the graphs, the maximum weight clique problem on the compatibility graph is solved. The maximum weight clique graph is a subgraph of G_c where all the vertices are pairwise adjacent and their total weight is maximum. Hence, the maximum weight clique graph will report the optimal node and edge merging, resulting in a maximum reduction of nodes (cost of processors), edges (cost of processor/FIFO ports) and edge capacities (size of FIFO buffers) in the merged graph. In Fig. 8.3f, the thick lines and thick bordered circles and ovals show the maximum weight clique graph for the running example (for the sake of simplicity, vertices' weights are omitted). Finding a maximum weight clique of a graph is known to be an NP-complete problem [21] and can be solved optimally using an exhaustive algorithm; however, the author used a polynomial-time algorithm from Cliquer tool [22] in the experiments. The resulting maximum weight clique is used to reconstruct G_{MG}. Firstly, the merged nodes and edges in G_{MG} are obtained from the clique. Then, all the nodes and edges in individual graphs that were not part of the clique are added to G_{MG}. Figure 8.3e shows the merged graph from MaxC which has only five nodes and five edges.

In MaxC, unlike MaxN, exhaustive permutations of merging all G_x is not required because the compatibility graph exhausts all possible merging of the two graphs. Hence, MaxC results in optimal G_{MG}; however, it will be exorbitantly slow for large graphs since the merging problem is NP-complete.

8.5 Experimental Methodology

Several benchmarks, reported in Table 8.1, consisting of hand-partitioned applications, StreamIt applications [23] and synthetic applications were used to create multi-mode pipelined MPSoCs. Hand-partitioned applications contain JPEG encoder, JPEG decoder and MP3 encoder. From StreamIt benchmark suite, the author chose Fast Fourier Transform (FFT), Beam Former (BF) and Time Delay Equalisation (TDE) applications. These are well-known streaming applications that appear frequently in embedded domain [1, 24]. Synthetic applications were used to evaluate the scalability of the heuristics with the increase in number of nodes and edges in application graphs.

The application graphs were merged using the three heuristics to derive a multi-mode pipelined MPSoC. The multi-mode pipelined MPSoC was created using Tensilica's Xtensa LX3 [25] processors with FIFO buffers between the processors. XTMP, ISS and XT-XENERGY tools were used to record the throughput, latency and energy consumption of the multi-mode pipelined MPSoC. For comparison of the three heuristics, the pipelined MPSoCs were not customised as customisation does not affect the merging of the application graphs. All the experiments were conducted on a 2.15 GHz quad core machine with 8 GB RAM.

Table 8.1 Benchmark characteristics

Application	# nodes	# edges
JPEGenc	7	9
JPEGdec	5	6
MP3enc	5	5
FFT	12	12
BF	12	12
TDE	13	12
Syn1	14	15
Syn2	14	15
Syn3	17	20

8.6 Results and Analyses

The results of merging different applications is shown in Table 8.2. The first column reports the merged applications. For example, the JPEGEnc/Dec/MP3Enc means that JPEGEnc, JPEGDec and MP3Enc were combined. The second and sixth minor columns, denoted as Ind., represent the traditional approach where individual application graphs are separately mapped to pipelined MPSoCs. Hence, the number of nodes and edges for Ind. would be the sum of the number of nodes and edges in individual application graphs. For example, the number of nodes in Ind. for JPEGEnc/Dec is 12 due to 7 and 5 nodes in JPEGEnc and JPEGDec respectively (from Table 8.1). The rest of the minor columns report the number of nodes and edges in merged graphs from MaxS, MaxN and MaxC. As expected, MaxC results in the least number of nodes and edges in all the merged application graphs. In some cases, MaxN results in higher number of edges than MaxS because it only focuses on reducing the number of nodes. For example, in JPEGEnc/MP3Enc, MaxN produced 11 edges compared to 10 from MaxS. The optimality of MaxC comes at the expense of higher running time which is reported in the fourth major column named Time. MaxS and MaxN heuristics take a few seconds to run while MaxC's running time is several minutes. Three synthetic benchmarks are used to stress the heuristics, and the results are reported in the last three rows. MaxC's running time scales poorly with the increase in number of nodes and edges as it did not finish in even 4 days for Syn1/Syn3 and Syn2/Syn3. Hence, for large graphs, MaxS and MaxN should be used instead.

Figure 8.4 shows the area saving for the first eight merged graphs from Table 8.2 (denoted (a) – (h)) compared to Ind. Synthetic benchmarks were not mapped to multi-mode pipelined MPSoCs. For each merged graph, the area saving is broken down in processor (P) and FIFO (F) area saving (in gates), plotted on the left y-axis as a percentage; and, the number of processor/FIFO ports (#P) saved, plotted on the right

Table 8.2 Comparison of MaxS, MaxN and MaxC heuristics

Merge	#nodes				#edges				Time		
	Ind.	MaxS	MaxN	MaxC	Ind.	MaxS	MaxN	MaxC	MaxS	MaxN	MaxC
JPEGEnc/Dec	12	9	7	7	14	12	12	8	<1s	<1s	1m
JPEGEnc/MP3Enc	12	8	7	7	13	10	11	8	<1s	<1s	1m
JPEGDec/MP3Enc	10	6	5	5	11	7	8	6	<1s	<1s	1m
JPEGEnc/Dec/MP3Enc	17	9	7	7	19	12	13	8	1s	2s	3m
FFT/BF	24	14	12	12	24	16	14	13	1s	1s	5m
FFT/TDE	26	18	13	13	25	18	22	14	1s	1s	5m
BF/TDE	26	17	13	13	25	17	20	14	1s	1s	5m
FFT/BF/TDE	38	19	13	13	37	21	23	15	1s	2s	12m
Syn1/Syn2	28	18	14	14	30	24	29	20	1s	1s	16h
Syn1/Syn3	31	21	17	N/A	35	25	32	N/A	1s	1s	>4d
Syn2/Syn3	31	26	17	N/A	35	23	31	N/A	1s	1s	>4d

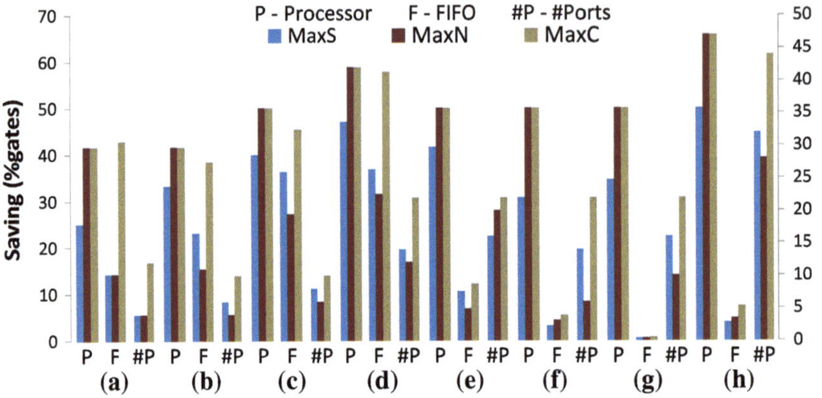

Fig. 8.4 Reduction in processor and FIFO area (*left y-axis*) and number of processor/FIFO ports (*right y-axis*)

y-axis. MaxN always saves the same amount of processor area as MaxC since it uses the maximum number of nodes from all the application graphs. However, MaxC saves more FIFO area and processor/FIFO ports. For example, in FFT/BF/TDE, MaxC saved 44 ports compared to 28 and 31 from MaxN and MaxS which is significant considering the cost of wires and interconnects in the ports. In summary, multi-mode pipelined MPSoCs saved up to 62 % processor area (FFT/BF/TDE), 57 % FIFO area (JPEGEnc/Dec/MP3Enc) and 44 processor/FIFO ports (FFT/BF/TDE) compared to individual pipelined MPSoCs.

The author also compared the performance of the applications executing on the multi-mode pipelined MPSoC with their individual counterparts. An average degradation of 1 % in throughput and 2 % in latency with a standard deviation of 1 % and 2 % respectively was observed. The energy consumption per iteration of the multi-mode pipelined MPSoC increased by a maximum of 3 % due to the degradation in throughput and latency. These results indicate that multi-mode pipelined MPSoCs can be used as an execution platform for multiple, mutually exclusive multimedia applications. In addition, multi-mode pipelined MPSoCs can be designed by merging application graphs using the proposed heuristics.

8.7 Discussion

Use of all the three heuristics for merging application graphs is not necessary. Ideally, a designer should use MaxC to find an optimal merging. When MaxC takes an exorbitant amount of time, the designer should utilise MaxS and MaxN to quickly gain knowledge of a possible merging. MaxS performs poorly in reducing the processor area (see Fig. 8.4) compared to MaxN which provides the same amount of processor area saving as MaxC. However, MaxS has a higher chance (six out of eight times in Fig. 8.4) of reducing processor/FIFO ports compared to MaxN since it works at graph topology level, although this cannot be proved because processor/FIFO port reduction depends on graph topologies, and both MaxS and MaxN are heuristics. A designer should first use MaxN to merge the application graphs, and then use MaxS to gain an insight into reducing the number of processor/FIFO ports. This is possible since both MaxS and MaxN are quite fast (see Table 8.2).

8.8 Summary

This chapter proposed multi-mode pipelined MPSoCs where one application is executed in a given mode. The author proposed merging of the application graphs into a single graph to design a multi-mode pipelined MPSoC. Application graphs are merged using two greedy heuristics (MaxS and MaxN) and a maximum weight clique based approach (MaxC) so that the number of nodes and edges with their capacities is minimal in the merged graph. The results show that multi-mode pipelined MPSoCs derived from merged graphs using MaxC save up to 62 % processor area, 57 % FIFO area and 44 processor/FIFO ports compared to individual pipelined MPSoCs. In all the multi-mode pipelined MPSoCs, minuscule degradation in throughput and latency and an increase in energy consumption per iteration was observed. These results indicate viability of multi-mode pipelined MPSoCs as multi-mode accelerators in a multimedia platform.

References

1. H. Javaid, A. Ignjatovic, S. Parameswaran, Rapid design space exploration of application specific heterogeneous pipelined multiprocessor systems. Trans. Comput. Aided Des. Integr. Circuits Syst. **29**, 1777–1789 (2010)
2. S.L. Shee, S. Parameswaran, DAC '07: Design methodology for pipelined heterogeneous multiprocessor system, in Proceedings of the 44th annual conference on Design automation, pp. 811–816, 2007
3. Texas Instruments, Omap mobile processors. Available at http://www.ti.com/
4. NVIDIA, Tegra multiprocessor architecture. Available at http://www.nvidia.com/
5. W. Geurts, Accelerator Data-Path Synthesis for High-Throughput Signal Processing Applications. Kluwer Academic Publishers, 1997
6. N. Shirazi, W. Luk, P. Cheung, Automating production of run-time reconfigurable designs, in *Proceedings of IEEE Symposium on FPGAs for Custom Computing Machines*, pp. 147–156, Apr 1998
7. Z. Huang, S. Malik, Managing dynamic reconfiguration overhead in systems-on-a-chip design using reconfigurable datapaths and optimized interconnection networks, in *Proceedings of Conference and Exhibition 2001 on Design, Automation and Test in Europe*, pp. 735–740, (2001)
8. N. Moreano, E. Borin, C. de Souza, G. Araujo, Efficient datapath merging for partially reconfigurable architectures. IEEE Trans. Comput. Aided Des. Integr. Circuits Syst. **24**, pp. 969–980, (2005)

9. Y. J. Chong S. Parameswaran, Custom floating-point unit generation for embedded systems. IEEE Trans. Comput. Aided Des. Integr. Circuits Syst. **28**, pp. 638–650, (2009)
10. P. Brisk, A. Kaplan, M. Sarrafzadeh, Area-efficient instruction set synthesis for reconfigurable system-on-chip designs, in *Proceedings of 41st Design Automation Conference*, pp. 395–400, (2004)
11. H. Oh, S. Ha, in CODES '02: Hardware-software cosynthesis of multi-mode multi-task embedded systems with real-time constraints, in *Proceedings of the Tenth International Symposium on Hardware/Software* Codesign, (2002)
12. V. Kianzad, S. Bhattacharyya, Charmed: a multi-objective co-synthesis framework for multi-mode embedded systems, in *Proceedings of 15th IEEE International Conference on Application-Specific Systems, Architectures and Processors*, pp. 28–40, (20040
13. M. Schmitz, B. Al-Hashimi, P. Eles, Cosynthesis of energy-efficient multimode embedded systems with consideration of mode-execution probabilities. IEEE Trans. Comput. Aided Des. Integr. Circuits Syst, **24**(2), 153–169 (2005)
14. M. Kim, S. Banerjee, N. Dutt, N. Venkatasubramanian, Energy-aware cosynthesis of real-time multimedia applications on mpsocs using heterogeneous scheduling policies, ACM Trans. Embed. Comput. Syst. **7**, pp. 9:1–9:19, (2008)
15. L. Huang, Q. Xu, Energy-efficient task allocation and scheduling for multi-mode mpsocs under lifetime reliability constraint, in *Proceedings of Design, Automation Test in Europe Conference Exhibition (DATE)2010*, pp. 1584–1589, Mar 2010
16. N. Moreano, G. Araujo, C. de Souza,Cdfg merging for reconfigurable architectures' Technical Report on IC-03-18, Institute of Computing UNICAMP (2003)
17. A. Kumar, S. Fernando, Y. Ha, B. Mesman, H. Corporaal, Multiprocessor systems synthesis for multiple use-cases of multiple applications on fpga, ACM Trans. Des. Autom. Electron. Syst. vol. **13**, pp. 40:1–40:27, (2008)
18. A. Shabbir, A. Kumar, S. Stuijk, B. Mesman, H. Corporaal, Ca-mpsoc: an automated design flow for predictable multi-processor architectures for multiple applications, J. Syst. Architect. (2010)
19. A. K. Singh, A. Kumar, T. Srikanthan CASES '11: A hybrid strategy for mapping multiple throughput-constrained applications on mpsocs, in *Proceedings of the 14th International Conference on Compilers, Architectures and Synthesis for Embedded Systems*, (New York, NY, USA), pp. 175–184, ACM, 2011
20. S. Wildermann, J. Angermeier, E. Sibirko, and J. Teich, Placing multimode streaming applications on dynamically partially reconfigurable architectures, Int. J. Reconfig.Comput. textbf2012, pp. 9:9, (2012)
21. E. Balas, V. Chvátal, J. Nešetřil, On the maximum weight clique problem, Math. Oper. Res. **12**, no. 3, pp. 522–535, (1987)
22. Cliquer. Available at http://users.tkk.fi/ pat/cliquer.html
23. W. Thies, M. Karczmarek, S. P. Amarasinghe, CC '02: Streamit: A language for streaming applications, in *Proceedings of the 11th International Conference on Compiler Construction*, pp. 179–196, (Springer-Verlag), 2002
24. M. Hashemi, S. Ghiasi, Throughput-driven synthesis of embedded software for pipelined execution on multicore architectures, ACM Trans. Embed. Comput. Syst. **8**, pp. 11:1–11:35, (2009)
25. Tensilica, Xtensa Customizable Processor. http://www.tensilica.com

Chapter 9
Conclusions and Future Work

This monograph explored implementation of multimedia applications on a pipelined MultiProcessor System on Chip (MPSoC) where the processors were divided into stages, which are connected in a pipeline. Application Specific Instruction set Processors (ASIPs) were used so that their customisation could be exploited to balance the workload across stages of the pipelined MPSoC; therefore, improving utilisation of the processors for high performance, reduced area footprint and low power consumption. Thus, each processor in the pipelined MPSoC had a number of configurations trading-off performance and area footprint, where one combination of processor configurations made up one of the pipelined MPSoCs design points. The aim of the monograph was to optimise such a pipelined MPSoC for the area footprint and energy consumption under performance constraints.

This monograph proposed design-time and run-time optimisations, which were targeted at different objective functions. Firstly, a pipelined MPSoC was optimised for the area footprint under either a latency constraint or a throughput constraint by selection of the most suitable processor configurations during its design space exploration. Then, such a design-time optimised pipelined MPSoC was augmented with run-time adaptability to deactivate idle processors or transition them to low-power states at run-time for low-power operation under a dynamic workload. Finally, the pipelined MPSoCs that had been optimised for different multimedia applications were combined into a single multi-mode pipelined MPSoC for further reduction of the area footprint. The proposed design-time and run-time optimisations have shown that pipelined MPSoCs can emerge as a viable implementation platform for multimedia applications. The following paragraphs summarise the proposed optimisations and the corresponding results.

Chapters 4 and 5 targeted quick design space exploration of pipelined MPSoCs for area footprint optimisation. Chapter 4 proposed analytical models to estimate the execution time, latency and throughput of a pipelined MPSoCs design point using latencies of individual processor configurations, and thus avoiding slow, full-system, cycle accurate simulations of all the design points. For effective use of these analytical models, latencies of individual processor configurations were gathered

H. Javaid and S. Parameswaran, *Pipelined Multiprocessor System-on-Chip for Multimedia*, DOI: 10.1007/978-3-319-01113-4_9,

with minimal number of simulations by utilising two estimation methods (PS and PSP). The PS method simulated all the processor configurations once, while the PSP method simulated only a subset of processor configurations and then used a processor analytical model to estimate the latencies of the processor configurations. Experiments with five pipelined MPSoCs executing typical multimedia applications (JPEG encoder/decoder, MP3 encoder and H.264 encoder) showed that the analytical models with PS and PSP methods had maximum absolute errors of 12.95 and 18.67 % respectively, and minimum fidelities of 0.93 and 0.88 respectively. Compared to the PS method, the PSP method reduced simulation time from days to several hours for design spaces that ranged from 10^{12} to 10^{18} design points.

Chapter 5 followed on from Chap. 4 by utilising the analytical models for quick design space exploration. Integer Linear Programming (ILP) formulations for area footprint optimisation under an execution time and a latency constraint, and an algorithm for area footprint optimisation under a throughput constraint were proposed. The proposed exploration techniques were evaluated using the five pipelined MPSoCs created in Chap. 4, which had design spaces up to 10^{18} design points. The time to find the Pareto front of each pipelined MPSoC with respect to latency or throughput was less than seven minutes, illustrating the applicability of the proposed design space exploration methods.

Next, in Chaps. 6 and 7, run-time optimisations were proposed to reduce energy consumption of a pipelined MPSoC. Chapter 6 proposed an adaptive pipelined MPSoC architecture, capable of adapting itself to run-time variations in its workload. In an adaptive pipelined MPSoC, stages with significant run-time variations in workload are implemented using *Main Processors* and *Auxiliary Processors*, where the main processor used differing numbers of auxiliary processors, considering the run-time workload variations. A main processor was equipped with a run-time processor manager which used a combination of the application's execution and knowledge (algorithmic and data properties) and information from the off-line profiling and statistical analysis to proactively predict the number of auxiliary processors that should be used. The idle auxiliary processors were either clock- or power-gated to reduce energy consumption. Experiments with an H.264 video encoder, designed for HD720p at 30 fps, showed that an adaptive pipelined MPSoC provided an energy reduction of up to 34 and 39 % for clock- and power-gating based deactivation of auxiliary processors respectively with a minimum throughput of 28.75 fps compared to a worst-case pipelined MPSoC.

Chapter 7 proposed a power manager where auxiliary processors had multiple power states, trading-off the overhead of the transition to power states with their possible energy reductions. Five heuristics were proposed as part of the power manager to forecast at run-time the duration of an upcoming idle period of an auxiliary processor using either the application's execution or the application's knowledge. Then, based on the predicted duration of the idle period, the most suitable power state was selected. Compared to the use of the processor manager with only clock-gating or only power-gating in an adaptive pipelined MPSoC executing H.264 video encoder (HD720p at 30 fps), the power manager reduced up to 40 % more energy with only an additional 0.5 % degradation of the throughput.

Finally, Chap. 8 proposed to create multi-mode pipelined MPSoCs by merging pipelined MPSoCs optimised for individual multimedia applications for further reduction of area footprint. To this end, individual application graphs were merged into a single graph by finding a maximal overlap between the graphs. Three heuristics were proposed where two of them greedily merged application graphs, while the third one found an optimal merging at the cost of higher running time. The results indicated significant area savings (up to 62 % processor area, 57 % FIFO area and 44 processor/FIFO ports) with minuscule degradation of the system throughput (up to 2 %) and latency (up to 2 %) and an increase in energy per iteration (up to 3 %) when compared to individual pipelined MPSoCs.

Future works on this monograph can be conducted in several directions. Design space exploration of a pipelined MPSoC can consider differing communication architectures in addition to differing processor configurations. For example, data transfers in a pipelined MPSoC can be achieved using dedicated FIFO buffers (as was done in this monograph), Direct Memory Access (DMA) engines or software managed FIFO buffers in a shared memory. Such an exploration will not only optimise the computational architecture (processors), but also the communication architecture of a pipelined MPSoC, resulting in better area footprint optimisation.

A pipelined MPSoC can use reconfigurable processors as its building blocks rather than Application Specific Instruction set Processors (ASIPs) to further reduce the area footprint and improve system adaptability. Reconfigurable processors with so-called reconfigurable regions can load custom instructions at run-time depending upon the needs of the sub-kernel, and thus can time-multiplex the reconfigurable regions for reduced area footprint. Furthermore, these reconfigurable regions can be turned off to reduce energy consumption if none of the custom instructions are required. However, the introduction of reconfigurable processors in pipelined MPSoCs will require run-time management techniques at the processor-level, in addition to the system-level techniques proposed in this monograph. The adaptability feature of reconfigurable processors can also be exploited in a multi-mode pipelined MPSoC where the reconfigurable regions are loaded with the only custom instructions required of the currently executing application.

The adaptability of the adaptive pipelined MPSoC proposed in this monograph was exploited to reduce energy consumption only. One of the future works can exploit the adaptability of an adaptive pipelined MPSoC for resource management, where auxiliary processors of one stage can be used as the auxiliary processors of an other stage depending upon the workload of the stages. In other words, a pool of auxiliary processors can be shared among multiple stages of the adaptive pipelined MPSoC, where allocation of auxiliary processors to a particular stage is done at run-time by the resource manager. This will require run-time resource management heuristics.

Lastly, an operating system can be designed for the pipelined MPSoC so as to manage its applications, resources and power consumption at run-time. Such an operating system can allow simultaneous execution of multiple applications by context switching between them in a multi-mode pipelined MPSoC. To this end, an efficient and fast context switch method will be required for not only the processors, but also the FIFO buffers between the processors.

Index

H. Javaid and S. Parameswaran, *Pipelined Multiprocessor System-on-Chip for Multimedia*, DOI: 10.1007/978-3-319-01113-4,